IMAGES
of America

DAVISVILLE AND THE SEABEES

The above flag detail is from the 70th Naval Construction Battalion, a Seabee unit that assembled for training at Camp Endicott, Rhode Island. Here, battalion members received their boot training, which commenced on December 18, 1942. Following advanced training, some of the men deployed overseas as members of special pontoon detachments, where they became noted for their participation in the landings at Sicily, Salerno, and Anzio. Others saw service in Africa and the Pacific. (Courtesy Arthur Johnson, MMS 3/c.)

IMAGES
of America

DAVISVILLE AND THE SEABEES

Walter K. Schroder
and Gloria A. Emma

ISBN 0-7385-0106-9

First printed in 1999.
Reprinted in 2002.

Published by Arcadia Publishing,
an imprint of Tempus Publishing, Inc.
2A Cumberland Street
Charleston, SC 29401

Printed in Great Britain.

For all general information contact Arcadia Publishing at:
Telephone 843-853-2070
Fax 843-853-0044
E-Mail sales@arcadiapublishing.com

For customer service and orders:
Toll-Free 1-888-313-2665

Visit us on the internet at http://www.arcadiapublishing.com

SONG OF THE SEABEES

We're the Seabees of the Navy
We can build and we can fight
We'll pave a way to victory
And guard it day and night
And we promise that we'll remember
The Seventh of December
We're the Seabees of the Navy
Bees of the Seven Seas!

Contents

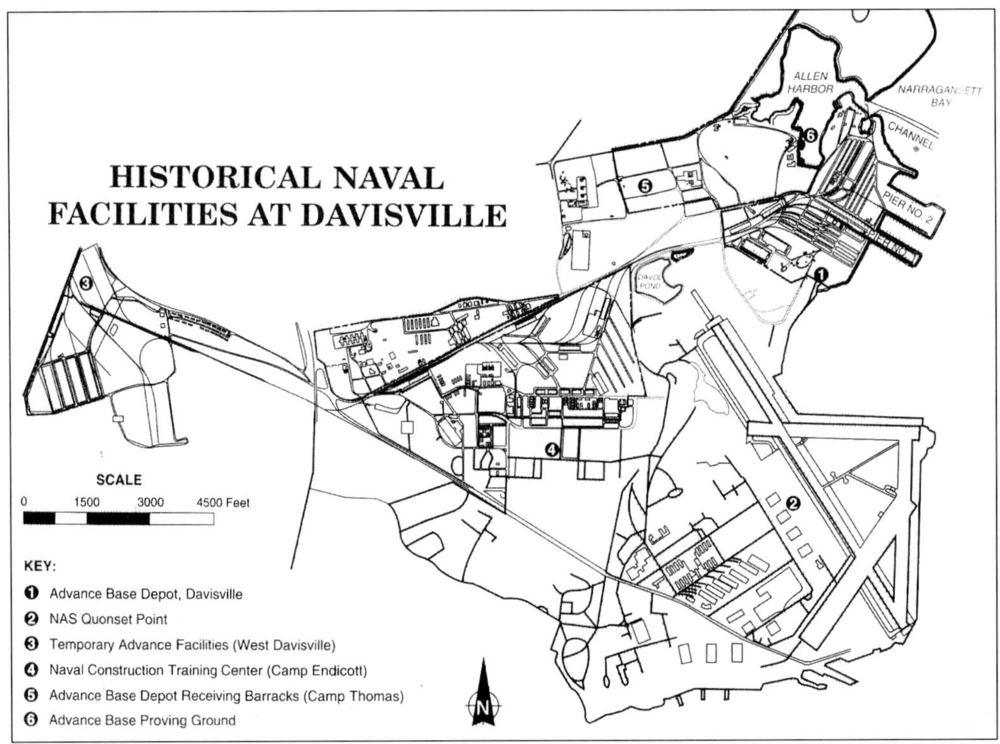

While the former naval facilities in North Kingstown, Rhode Island, are often viewed as a single entity and referred to broadly by the public as Quonset or Quonset Point, it is essential to note that within this enclave, there existed two distinct military entities, each with its own mission and responsibilities. Quonset Point was an U.S. Naval Air Station; Davisville was a Seabee Center, in fact it was the original home of the Atlantic Seabees. Review of the above map shows that the Naval Construction Battalion Center, Davisville, was the larger of the two in terms of acreage occupied, having encompassed the various components identified under key items one, three, four, five, and six. (NCBC Davisville Historical Perspective 1942–1994.)

Acknowledgments

The authors wish to thank all former Officers, Seabees, Civilian Employees, their families, and their friends, who contributed information and photographs for inclusion in this volume. Without their enthusiastic support, this book could not have been written. In addition, the authors extend their sincere thanks to the following organizations for their assistance in providing information and photographic images essential to the project: Navy Seabee Veterans of America, Island X-1, Davisville; the Rhode Island Economic Development Corporation, Davisville; and the on-site representative of Northern Division, Naval Facilities Engineering Command.

Special thanks also to Robert H. Emma, for scanning many of the images received; to Lora C. Schroder, for her help in the cataloging and layout phases of the project; and to Edward L. Turner, EOCS and Secretary, Navy Seabee Veterans, Island X1-Davisville, for his extraordinary personal efforts in support of this venture.

Introduction

The U.S. Naval Construction Battalion Center at Davisville, Rhode Island, is first and foremost remembered as the "Original Home of the Atlantic Seabees." It is here at this North Kingstown facility that 100 battalions and 56 other units of U.S. Navy Builder-Fighters were formed, outfitted, trained, and prepared for overseas deployment during World War II (WWII).

The Seabee base is located adjacent to the former Naval Air Station (NAS) Quonset Point, for which construction originally commenced in July 1940. In addition to its regular responsibilities, the 1,192-acre Quonset facility was also charged with the assembly of materials needed to set up several military bases in Europe under the 1941 Lend-Lease Program. As an outgrowth of this tasking, a tract of 85 acres was acquired at West Davisville for the construction of a facility to produce temporary shelters—the famous "Quonset Huts"—and to store related manufacturing components. West Davisville was then referred to as the Temporary Advance Facilities (TAF). As a necessity, crews were then required to erect the huts. An initial contingent of 258 naval construction personnel was trained at Quonset in early 1942 for that purpose and shipped overseas.

By the end of February 1942, the existing storage areas and facilities were determined insufficient to cope with increasing military demands. Added capacity became available by the establishment of the Advance Base Depot (ABD) Davisville, which became fully operational on 1,892 acres of land in June of 1942. Its mission included the manufacture, storage, and embarkation of materials required for advance base construction; the receiving and housing of troops; the training of personnel in construction and military skills; and the development and testing of special equipment. To comply with this tasking, a 142-acre personnel receiving station known as Camp Thomas was established under the ABD in October 1942. This area contained 500 Quonset huts for troop housing. On August 11, 1942, the Naval Construction Training Center (NCTC) Davisville was commissioned. Known as Camp Endicott, this encampment for 14,000 men had Seabee "training" as its primary mission. One of the last activities added to this growing military complex at Davisville was the establishment of the Advance Base Proving Ground on 23 acres of land in Allen Harbor. This activity engaged in pontoon development and experimentation.

The above Davisville facilities were fully operational for the duration of WWII as is further described in this volume. They came into being because of wartime necessity and the foresight of Rear Admiral (R.Adm.) Ben Moreell, then Chief of the Navy's Bureau of Yards and Docks (BuDocks). Late in December 1941, R.Adm. Moreell requested permission to recruit personnel skilled in essential construction trades so he could form a force of trained Naval Construction Battalions, who would be able to build and fight in the face of the enemy. He was granted the necessary authority in January 1942 and soon thereafter permission to call the members of the Construction Battalions "CBs" and to identify them and their units with the distinctive insignia of a fighting Bee. The motto given these units was "Construimus Batuimus"—WE BUILD, WE FIGHT! Over 100,000 Seabees were trained at Davisville's Camp Endicott during WWII.

On December 31, 1945, the Advance Base Depot and the Naval Construction Training Center at Davisville were disestablished. Only a Civil Engineer Corps Officers School remained at Davisville; however, by the beginning of September 1946, it was moved to another location.

Davisville experienced revitalization beginning on August 8, 1951, when, in the face of the Korean War, the Naval Construction Battalion Center, Davisville, was established. During the following 5 years, many new buildings and facilities were added; the base flourished once more with the new generation of Seabees on board, who were ready to serve their country wherever needed during the unsettled Cold War period. They, the Bees who preceded them in WWII, as well as the many dedicated civilian workers who supported them, deserve a hearty "well done." So profound is their combined story that the authors regret being able only to scratch the surface in retelling the story of Davisville and the Seabees in this single volume.

It is hoped that this modest presentation will give the reader a passing insight into the lives and deeds of the men and women who served Davisville's Naval Construction Forces and the civilian contingent that supported them faithfully until the base closed in 1994.

Postscript

The pace of the responses to the authors' requests for photographs and cruise books picked up considerably during the last weeks before the project deadline. This entailed the establishment of a cutoff date for submission of illustrations that could be considered for incorporation into the project.

In so doing and after evaluating the available photographs and cruise books, it became apparent that many more Seabee units had been stationed at Davisville during WWII than those represented in the information at hand. Because of this, the information contained in this volume is of necessity not all-inclusive. It is, however, representative of the many fine deeds and accomplishments of all Seabees that ever served at Davisville.

One

Advance Base Depot: The Beginning

The future rail spur from Devil's Foot Rock, North Kingstown (in the rear at Route 1), to Quonset Point was being cut through properties taken by the federal government. Initial construction got underway during the latter part of 1940. (Rhode Island Economic Development Corporation.)

L. Morra's vegetable and fruit stand at Devil's Foot Rock was immediately to the south of the new rail spur that would eventually cross Route 1 at this location. Construction of an overpass required the removal of this roadside business. (Rhode Island Economic Development Corporation.)

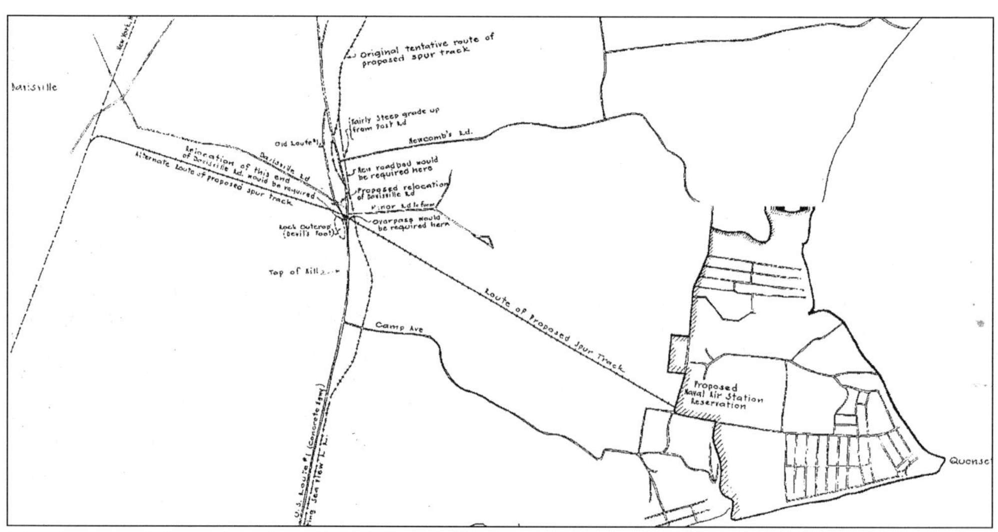

The layout of the rail and roadwork near Devil's Foot Rock is shown in this early plan for the construction of Quonset Point Naval Air Station. Future Davisville facilities would come into being soon after, in the area bordered by the proposed spur track and Newcomb Road. (Rhode Island Economic Development Corporation.)

This northward-looking image shows the planned spur crossing from 250 feet south of the site on Route 1. Romano Vineyards, an enterprise of some 370 acres that produced 100,000 gallons of wine per year from 125,000 grape vines, was located just north of the crossing on the right side of the road. It, too, had to give way to the needs of the military in 1940 and was eventually absorbed by the Davisville Seabee complex. (Rhode Island Economic Development Corporation.)

The F.D. Timson property across Route 1 at Devil's Foot Rock was in the path of the contemplated rail spur to Quonset Point and was condemned as part of the military construction project in September 1940. (Rhode Island Economic Development Corporation.)

The Quonset Point construction project, and the subsequent addition of Davisville to the Navy's plans, required employment of scores of able-bodied workers that were brought to their work sites via direct rail and bus transportation from Providence. (Rhode Island Economic Development Corporation.)

A virtual army of able-bodied workers was needed to build the Quonset and Davisville facilities. To facilitate recruitment and employment of the giant work force, it became necessary to set up worker registration offices in trailers near the work sites. In addition to normal registration procedures, the workers were photographed and fingerprinted for security and identification purposes. (Rhode Island Economic Development Corporation.)

Some of the first buildings at the Advance Base Depot were erected in 1941. Here, construction of the framework for a classroom building can be seen. Early contingents of Seabees underwent specialty training here, as well as in similar buildings that popped up overnight because of the urgency attached to the Quonset-Davisville construction projects. (Navy Photograph in Walter K. Schroder Collection.)

This building was tagged to become a garage at the Advance Base Depot. Built in 1941, the structure became one of many new facilities that rose out of the ground at record speed as a result of the urgency attached to the overall construction program. (Navy Photograph in Walter K. Schroder Collection.)

In 1941, the Temporary Advance Facilities, also referred to as West Davisville, was acquired as a manufacturing and storage plant for huts and associated components. Here, the loading of such materials at the Quonset Point pier can be observed. From this site, complete prefabricated facilities were shipped to remote overseas destinations, where Seabees would receive the materials and set up camps and other required facilities. (Navy Photograph in Walter K. Schroder Collection.)

A sizable piece of real estate was required to accommodate the Pontoon Storage Area of the Advance Base Depot and its three rail spurs. Viewing this aerial photograph brings to mind the WWII phrase, that America is the "Arsenal of Democracy." (Courtesy CWO3 Jack Sprengel, CEC, USN, retired.)

Pontoons developed, tested, and manufactured at West Davisville, beginning in 1941, literally paved the way for important Allied invasions at landing sites throughout Europe and the Pacific during WWII. (NCBC Davisville Historical Perspective 1942–1994.)

The Advance Base Proving Ground was established at Davisville in the spring of 1943. The facility was charged with the development of pontoon gear that was later used in landing operations in Europe and in the Pacific. Various configurations and uses of pontoons were tested here; designs acceptable to American, as well as Allied, forces were produced for their use under combat conditions. (Courtesy CWO3 Jack Sprengel, CEC, USN, retired.)

Because of his foresight and in recognition of his achievements during WWII, Adm. Ben Moreell became the youngest Vice Admiral in the U.S. Navy. While a Rear Admiral and Chief of the Bureau of Yards and Docks in 1941, he saw the need for organizing Naval Construction Battalions under the command of Civil Engineer Corps officers who could quickly be deployed to construct essential facilities while defending themselves under enemy fire. He submitted his request and plans on December 28, 1941, and received authority on January 5, 1942, to recruit men from construction trades for assignment to Navy construction units. He became known as the Navy's first "King Bee." (Courtesy Comdr. John Seites, CEC, USN, retired.)

The men of the 30th Naval Construction Battalion called Camp Thomas their home during the summer of 1944. While at the camp, they graded 10 acres of open storage area and laid a half-mile of spur line to facilitate the handling of supplies. Camp Thomas had movies, stage shows, a well-equipped gym, one of the country's largest swimming pools, a spacious bowling alley, a billiard room, and a cafeteria. Also of note was a Howard Johnson cafeteria near the main gate. The unit departed September 7, 1944; the last person they waved goodbye to was a uniformed guard at the spur gate—her own sons were "over there." (Navy Seabee Veterans of America, Island X-1, Davisville, Rhode Island.)

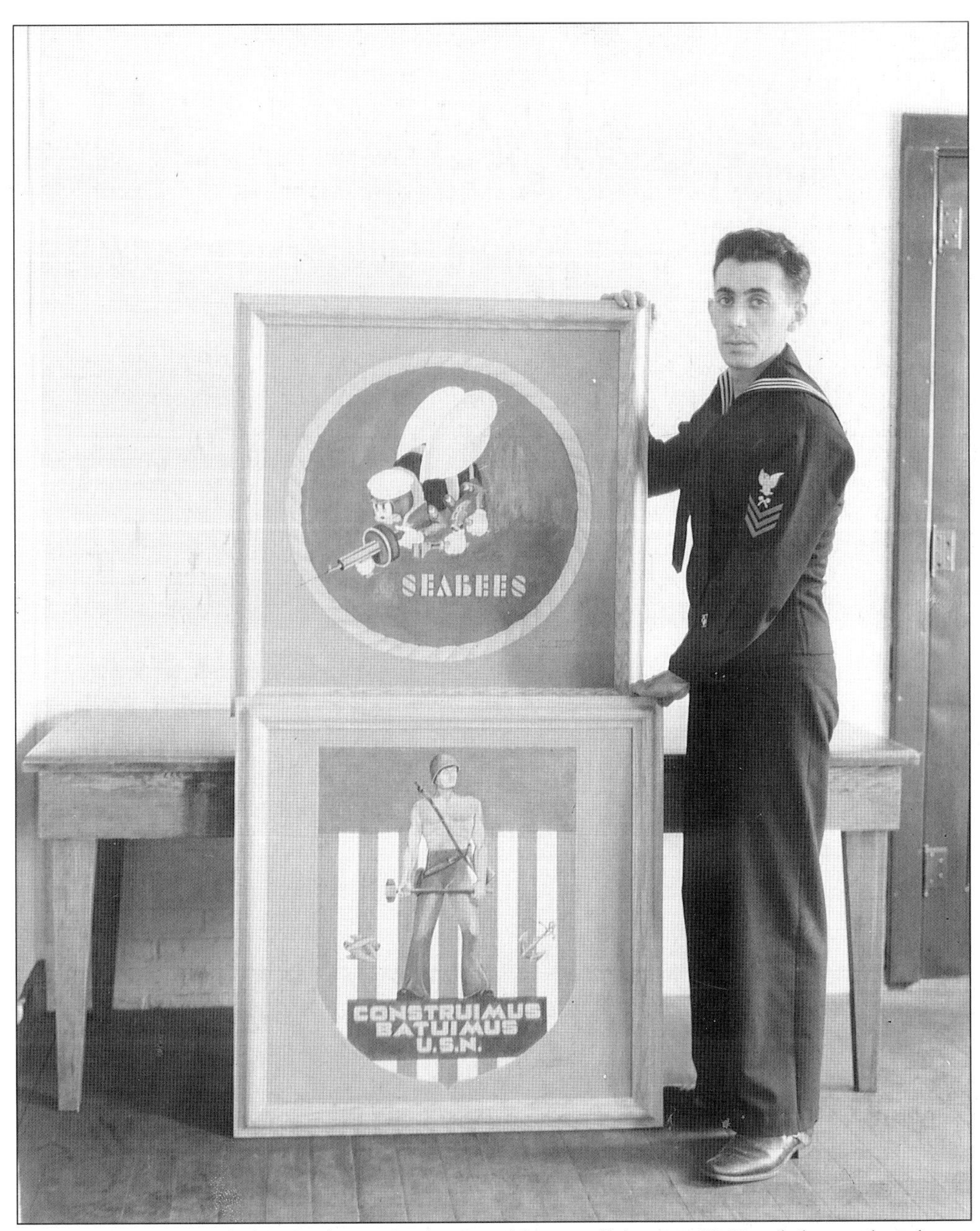

Frank Iafrate created the "Fighting Seabee" emblem in February 1942, while employed as a civilian at Quonset Point. Because of his talent for caricatures, he was asked to design an insignia for a special group of Navy men being trained in both construction and military skills. He considered several ideas, but centered his efforts on designing a "busy bee," one that works industriously and can fight to protect itself. The name "Seabee"—phonetically identical to CB (short for Construction Battalion)—was his choice for the insignia; he produced a fierce looking bee wearing a sailor's hat and carrying the tools and weapons of the trade. Iafrate became fascinated with the Seabees and signed up with them. Here, he is seen with renderings of his Seabee design and R.Adm. Ben Moreell's "Construimus Batuimus" motto. (Courtesy Frank J. Iafrate.)

In 1943, local volunteer hostesses formed a "CB-ettes Club" at the Advanced Base Depot, Davisville. They are gathered here in the Recreation Hall. These volunteers helped the young Seabees, away from home, with some of their routine administrative chores, including letter writing and setting long distance phone calls up for them. (Courtesy Louise Allan.)

Towlines, published twice a month by the Welfare and Recreation Department of Camp Thomas, provided new arrivals with essential information on the facilities at the camp as well as items of general interest to most military personnel. The men were eager to receive copies of *Towlines*, as is evident from communications reproduced in the "Letters to the Editor" section of the paper. The August 1944 issue of *Towlines* published two such letters, one from a member of CBMU 611, then in Oran, Algeria, and the other from a Seabee, transferred to Port Hueneme, California. Both praised the editorial staff of the paper. (Collection of Walter K. Schroder.)

In the summer of 1944, *Towlines* published this sketch of the Camp Thomas Chapel, a large 100-by-40-foot Quonset Hut-type warehouse that had been converted to serve the religious needs of assigned service personnel. The facility had a seating capacity of 320 and was open to and used by members of various religious faiths. (Collection of Walter K. Schroder.)

Camp Endicott, the Seabee training camp in North Kingstown, Rhode Island, during WWII, was named in honor of R.Adm. Mordecai Thomas Endicott, Civil Engineer Corps, U.S. Navy, who had been the Chief of the Navy's Bureau of Yards and Docks from 1898 to 1907. This likeness was made by Frank Iafrate, the man known for his design of the famous Seabee emblem. (Courtesy Frank J. Iafrate.)

Many a WWII Seabee saw this Camp Endicott Seabee and Shield. It is said that this camp offered one of the most spectacular and exacting training programs in the U.S. Navy. Recruits, age 18 to 38, came from many construction trades. They learned to adapt their skills to the needs of the Navy, while at the same time acquiring the fighting techniques to "dish it out" in the best commando and guerilla style. (Courtesy Mrs. Elbert S. Buch.)

The formal dedication of Camp Endicott took place on June 27, 1942, although some men had commenced their training prior to that date. Navy Secretary Frank Knox was on hand for the dedication ceremony. The capacity of 15,000 men and 350 officers was reached in November of that year. (Navy Photograph in Walter K. Schroder Collection.)

The Regimental Headquarters of the Naval Construction Training Center, Camp Endicott, Davisville, Rhode Island, had overall responsibility for the operation of various schools, including the training of signalmen, firefighters, carpenters, photographers, sub-grade construction workers, heavy equipment mechanics, steam fitters, refrigeration specialists, and even some divers. Ship-fitting, welding, demolition, dock building, carpentry, and salvage were additional types of training offered at Camp Endicott. (Courtesy CWO3 Jack Sprengel, CEC, USN, retired.)

The Camp Endicott Barracks housed the Seabee trainees and the units getting ready for overseas duty. While learning and practicing their professional specialties at Davisville, the Seabees attended Judo classes and were required to participate in close order drill—the manual of arms—and to learn the use of various weapons. (Courtesy CWO3 Jack Sprengel, CEC, USN, retired.)

This mural titled *Island Xtasy* by Frank Iafrate, designer of the famous Seabee emblem, drew some vivid comments from the men at Camp Endicott. A pamphlet of the day puts it this way: "We'll soon be shoving off for 'Island X,' which is a mythical zone of duty to which all 'Seabees' will soon be assigned. As seen through the eyes of our artist, who painted a mural of it in the Officer's Club, 'Island X' is a heavenly tropical isle inhabited by beautiful native girls who answer every sailor's beckon or wish. Its chief resources are gin wells, beer springs, and ice cream plants. The only work a sailor has to do is sit around and be waited on." How disappointing their assignment to the Pacific would be. (Courtesy Frank J. Iafrate.)

Of necessity to the operation of Camp Endicott was a power plant of sufficient capacity to support the growing Seabee encampment. Here is an inside view of the boiler room, which was housed in a 40-by-100-foot Quonset Hut. (Courtesy CWO3 Jack Sprengel, CEC, USN, retired.)

The various types of facilities being constructed for the Seabees at Camp Endicott included a hospital and other buildings dedicated to the health and welfare of the Navy population. This picture captures this complex with the Isolation Ward in the foreground. (Rhode Island Economic Development Corporation.)

Described as one of the largest inside swimming pools at the time, this Camp Endicott facility was built early in 1944. The above photograph was taken on May 6, 1944. (Rhode Island Economic Development Corporation.)

"Look Ma, no hands." Seabees and sailors must learn to swim and stay afloat as their duties may require working in deep water. As Navy men, swimming is among the many skills they must master before going to sea. These men are going through a prescribed drill at the Camp Endicott swimming pool. (Collection of Virginia Dulleba.)

Training at Camp Endicott included diverse courses of instruction. Here, a Seabee operates a Whale Boat on Narragansett Bay as part of his preparation for an overseas assignment with one of the Davisville units. From his demeanor, he appears confident in his performance. (Courtesy Louise Allan.)

Bayonet training is a skill essential in hand-to-hand combat when an enemy has to be driven from his position by men on foot. Camp Endicott instructors, who at first seemed to be pretty human fellows, turned into fierce tigers charging their prey relentlessly during practice periods, as depicted here. (Collection of Virginia Dulleba.)

Over the side and down, negotiating the landing nets was a skill Seabees acquired at Camp Endicott. There was little doubt they would be required to put this training to good work at landing and invasion sites overseas. (Collection of Virginia Dulleba.)

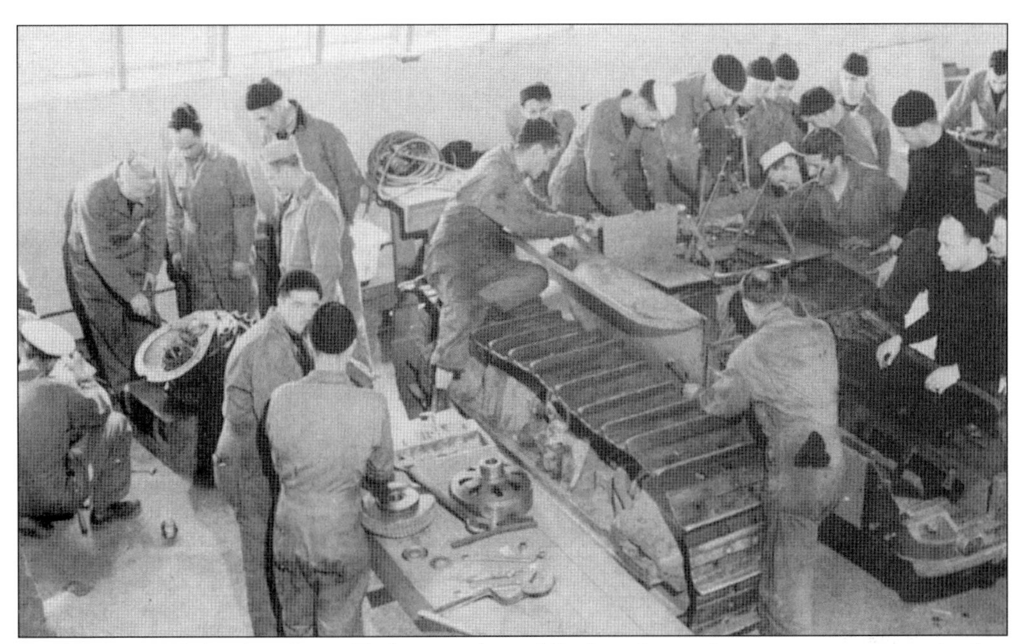

At the Engine School at Camp Endicott, selected Seabees learned to repair tractors and other earth-moving equipment of the types that would be assigned to their units in overseas areas. Their skills would be depended on to find and fix any equipment problems once they graduated and were no longer at their home base in Davisville. (NCBC Davisville Historical Perspective 1942–1994.)

To facilitate the training of stevedores, a life-size, mock Liberty ship was built at another site and moved to ABD Davisville, where special cargo-handling units of Seabees utilized the on-board loading equipment and cranes to perfect their skills on this shore-borne ship. (Courtesy CWO3 Jack Sprengel, CEC, USN, retired.)

Signaling with flags has long been a Navy specialty. The Camp Endicott curriculum was certain to include training for this unique skill. Here, on March 24, 1944, a class of Seabees practices various types of signaling. (Courtesy Mrs. Elbert S. Buch.)

In May 1944, Electricians Mate First Class Elbert S. Buch is shown instructing a group of Seabees in the art of signaling. The instructor eventually deployed to Okinawa with Construction Battalion Maintenance Unit 617 (CBMU-617). (Courtesy Mrs. Elbert S. Buch.)

These Seabees are attending a lecture at the Truck Driving School of the Naval Construction Training Center (NCTC), Camp Endicott, in February 1945. The markings on the backs of the outerwear worn by some of the trainees imply that the jackets have been issued by the local NCTC Supply Room. (Navy Photograph in Walter K. Schroder Collection.)

This group of 12 military and civilian personnel are posing during a break in their Heavy Equipment Repair Training class, taken sometime during the 1944–45 period. Every attempt was made to keep the classes to a manageable size. (Courtesy Edward Luth, MM2c.)

Several Royal Dutch Marines pose at Camp Endicott in October 1945, where they were attending a Heavy Equipment School session conducted by Ed Luth. This class was the last offering at the Heavy Equipment School before it was discontinued. (Courtesy Edward Luth, MM2c.)

When local Seabee designer Frank Iafrate was transferred to Port Hueneme, he had an opportunity to experience the site preparation for an airfield firsthand. This picture illustrates several of the special construction skills that were being taught at Camp Endicott in Rhode Island. (Courtesy Frank J. Iafrate.)

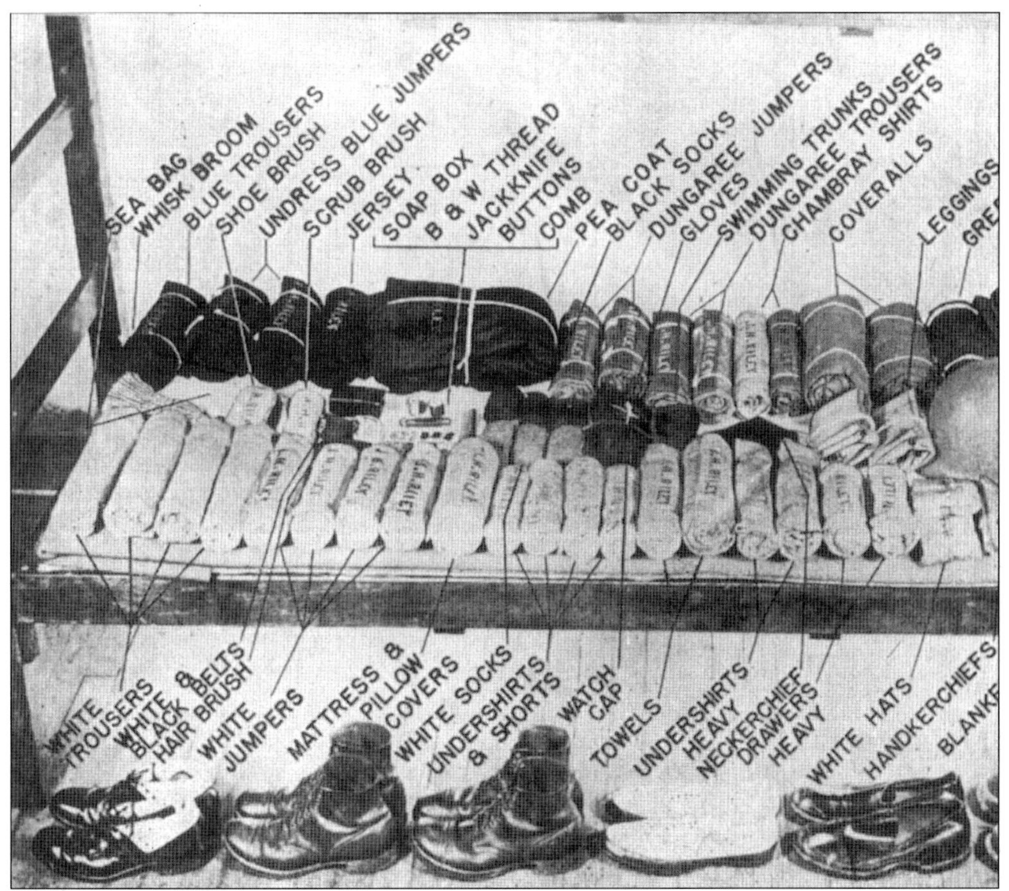

Each Navy Seabee was taught to follow the above illustration when displaying the contents of his duffel bag for inspection. It is amazing that all items shown could fit into a single duffel bag, considering that 6 pairs of footwear, a helmet, and several other bulky items were included in the personal clothing allowance. (Courtesy CWO3 Jack Sprengel, CEC, USN, retired.)

In September 1944, 'D' Co. of the 70th Naval Construction Battalion at Davisville returned from North Africa, where their ships came under attack by submarines while en route to Bizerta. Several men were lost and many others wounded. In Bizerta, they came under attack by German planes; however, the 70th always gave a good account of itself, no matter where they were assigned. (Courtesy Arthur Johnson, MMS 3/c.)

"Pass and Review" can be seen at Camp Endicott. Here, the 1006th Naval Construction Battalion Detachment (CBD) follows a unit of WAVES (Women Accepted for Volunteer Emergency Service) as they pass the reviewing stand. The 1006th became known for their participation in the landings at Sicily, Salerno, and Anzio. (Courtesy Ada Aren.)

This photograph shows inspection time for the WAVES Detachment at Camp Endicott. The WAVES served the installation and the Seabees admirably by performing many essential administrative and personnel duty assignments, thus freeing up men for overseas and combat duty. (Collection of Virginia Dulleba.)

From 1945 to 1946, Ella Matteson of Hamilton, Rhode Island, served as a civilian guard at the Quonset-Davisville Gates. She was about 50 years old at the time. Departing Seabees of the 30th Naval Construction Battalion noted in their logbook that the Civilian Guards were the last people from Davisville to whom they waved goodbye. (Courtesy Don Spink, grandson.)

A proud Seabee returns home after the war. John Doty of Jamestown, Rhode Island, shows off his distinctive Seabee shoulder patch as he poses during the summer of 1945, following his return home from overseas duty. (Courtesy John Doty Jr.)

The 137th Construction Battalion was one of many units that trained at Camp Endicott before being shipped overseas during WWII. This unit was assigned to Okinawa, a site many Davisville Seabees became familiar with during their Pacific tour of duty. The men are wearing their "blues," an indication that this photograph was taken during a cool time of year. (Collection of Gloria A. Emma.)

Another Seabee construction battalion parades proudly at Camp Endicott while the band strikes up a stirring marching tune, known to help keep the men in step. In passing the reviewing stand, they dip their unit flag out of respect for the reviewing party. However, the Stars and Stripes is held high above all. (Collection of Virginia Dulleba.)

This aerial view shows the Temporary Advance Facility commonly known as West Davisville. The buildings of the Assembly and Fabricating Plant for the now famous "Quonset Huts" can readily be seen in the center just above the rail line to Quonset Point, situated in the upper right corner of the picture. (Rhode Island Economic Development Corporation.)

After WWII, the West Davisville area was used as a storage site for Strategic War Reserves, which included raw materials essential to the National Defense, in particular types that were in critical or short supply. A 60-foot-high steel, elevated guard tower was erected at that time as part of an enhanced perimeter security system. (NORTHDIV, Naval Facilities Engineering Command.)

This view of the storage area at Plant No. 1 of the West Davisville Assembly and Fabrication facility was taken in 1941. Materials and hut components assembled here were moved to the Quonset Point pier by rail and shipped overseas from there. During WWII, this facility employed 3,000 workers who produced a total of 32,253 Quonset Huts. (Navy Photograph in Walter K. Schroder Collection.)

This string of Quonset Huts was at a training site on No Mans Island off Martha's Vineyard. The huts were typical of those produced at West Davisville during the war. They gave the appearance of semicircular arches in cross-section that had been joined together, much like a modified version of an older British Nisson hut. Due to the relatively simple design and adaptability of the huts, they were shipped to the most remote areas of the world during the war. (Navy Photograph in Walter K. Schroder Collection.)

Two

World War II Deployments

Men of Construction Battalion Maintenance Unit 617 (CBMU), originally from Davisville, pose in Okinawa in 1944. This detail operated an electrical shop, which is partially set up in the tent behind them. The group maintained the work done by others before them, made repairs, and finished any leftover projects. (Courtesy Mrs. Elbert S. Buch.)

Marines and Seabees of the 70th Naval Construction Battalion join hands to put logs and trees in place for a bridge across a jungle stream on New Britain Island. This cooperation is typical of that demonstrated under field conditions among Seabees and Marines. (Courtesy Arthur Johnson, MMS 3/c.)

Seabees of the 30th Naval Construction Regiment sort through a pile of supplies following their landing on an island in the Pacific. Markings on the crates identify Advance Base Depots Davisville and Port Hueneme as the originators of this shipment, which was necessary to the success of the assigned mission. (Navy Seabee Veterans of America, Island X-1, Davisville, Rhode Island.)

Men of the 30th Naval Construction Battalion can be seen erecting a 40-by-100-foot Quonset Hut on small Calicoan Island, located in the Philippines, in April 1945. The hut would become a galley, with a mess hall and scullery being built nearby. Because of the heat in the middle of the day, it was necessary to pour the concrete deck at night. (Navy Seabee Veterans of America, Island X-1, Davisville, Rhode Island.)

In 1945, the 101st Battalion was charged with a mass production effort to erect a series of storage huts in the Pacific area that required the use of all time-saving schemes the Seabees could think. It was said then that the "can do" motto allows for certain improvisations, as long as the end product serves its intended purpose. The job got done. (Barbara Smith in memory of Edward Fitzgerald.)

Despite the lack of privacy, the boys away from home, whether in or out of combat zones, were always anxious to keep in touch with their families. Thus, whenever there was a spare moment, it was used to write—or to play cards. (Barbara Smith in memory of Edward Fitzgerald.)

More important than writing letters was (and still is) the receipt of mail and packages from back home. These Seabees assigned to the 101st Naval Construction Battalion, somewhere in the Pacific in 1945, couldn't wait to get their hands on the goodies. (Barbara Smith in memory of Edward Fitzgerald.)

Here are typical WWII Seabees—hardworking, but never afraid to clown around or have a good laugh. These busy bees of the 70th NCB were on Iwo Jima near the now-famous Mount Suribachi when this picture was taken. (Courtesy Comdr. John Seites, CEC, USN, retired.)

The 62nd Naval Construction Battalion served at Oahu, Maui, and on Iwo Jima after training at Davisville in early 1943. Here, members of Headquarters Company pose with their carbines at one of the overseas stations. (Courtesy Louise Allan.)

Marines and Seabees supplemented each other, in particular in the Pacific battle zones, each providing the other with the expertise they had to offer. The Bees did the building while the Marines protected them to the best of their ability, so the builders could get the job done. At times, this special camaraderie went beyond normal bounds. Here, a Marine Corps Unit of Corsair fighter planes displays their adaptation of the fighting Seabee. (Courtesy Robert Mellor.)

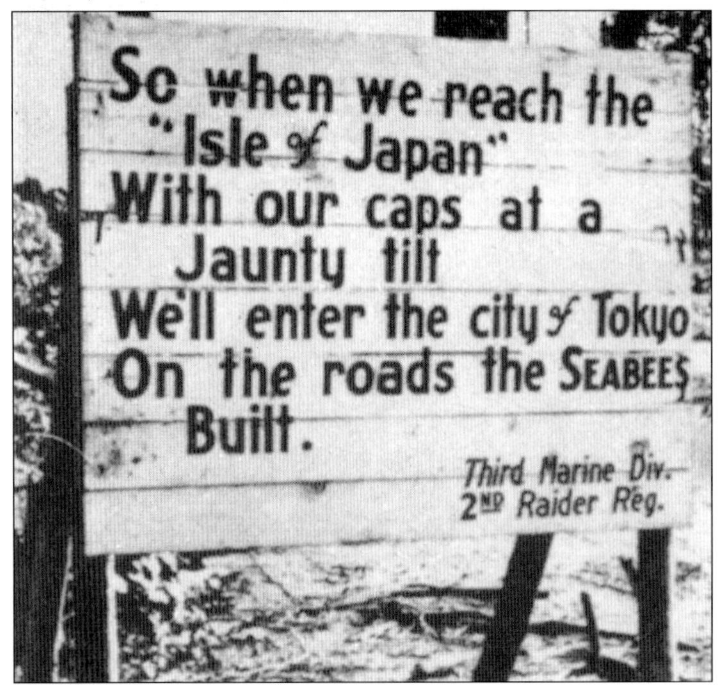

This third Marine sign posted somewhere in the Pacific says it all. The Seabees were essential and truly worth their salt. They could make things happen. As Gen. Douglas MacArthur put it, "The only trouble with the Seabees is that we don't have enough of them." (Courtesy Capt. A. Robert Socha, CEC, USN, retired.)

The Seabees were always sure to build a chapel wherever they were assigned. Here, a service is being conducted at a Seabee encampment somewhere in a remote area of the Pacific during WWII. (Courtesy Frank J. Iafrate.)

As taps are sounded, Seabees of the 133th Naval Construction Battalion bury their dead on Iwo Jima in 1945. The burial site is within view of Mount Suribachi and close to a larger cemetery, which can be seen to the rear of the men. (NCBC Davisville Historical Perspective 1942–1994.)

Having attended their first church service in the Pacific in an open field, the men of the 101st NCB went to work to build a chapel. They completed it in no time and took pride in their accomplishment. It became their spiritual home for months. For Protestants, there was a Candlelight Service on Christmas Eve and for Catholics a Holy Mass on Christmas morning. (Barbara Smith in memory of Edward Fitzgerald.)

During their assignment to North Africa in WWII, the 70th Naval Construction Battalion assisted the 120th Naval Construction Battalion in the construction of a hospital complex at Oran, as well as with other projects. When circumstances allowed, some of the men attended Sunday services while others lead the hymn, sung, or played the field organ. (Courtesy Arthur Johnson, MMS 3/c.)

The 70th NCB was instrumental in completing the above encampment in North Africa during WWII. Note the more permanent structures in the background and the large tent encampment in the foreground. Site preparation was one of the main tasks necessary to the success of the project. (Courtesy Arthur Johnson, MMS 3/c.)

Navy Seabees of the 70th NCB celebrate Armistice Day in Arzew, North Africa, with a parade. After the official function of the day, the men were granted passes to visit some of the sites nearby. (Courtesy Arthur Johnson, MMS 3/c.)

After the formation of the 1006th NCB detachment as a specialized pontoon unit, the outfit was deployed to such sites as Algeria, England, Normandy, and Sicily. While in England, in 1944, the unit was standing by at the Royal Naval College in Dartmouth, just prior to the Normandy invasion. (Courtesy Ada Aren.)

The 1006th NCB earned special fame during the landings at Salerno, Italy, in WWII. By then the men had been given the name "Pontoon Vets," especially after 10,000 personnel and equipment rolled over the floating pontoon bridges they had assembled. It is comforting to know that these Seabees had trained and learned their skills at Camp Endicott and the Advance Base Proving Ground, in Davisville, Rhode Island. (Courtesy Ada Aren.)

Some of the 1006th's Pontoon Vets are shown here at an undisclosed site. They appear pleased with themselves, as they should be, for the Pontoon Bees did a remarkable and praiseworthy job wherever they were sent. (Courtesy Capt. A. Robert Socha, CEC, USN, retired.)

This picture, taken on July 24, 1943, in Gela, Sicily, shows four men of the 1006th NCB posing after a successful landing. Counterclockwise, from front left, are Don Johnson, Mike Clements, Leo Cyre, and Don Arentowicz. (Courtesy Ada Aren.)

When the war ended, the Davisville facilities were quickly disestablished. For a period of time, the only organization remaining was the Civil Engineer Corps Officers School (CECOS) that had been established on May 5, 1945. It continued to operate until September 1, 1945, at which time the function was transferred to Port Hueneme, California. The above photograph shows the Ships Company of the Officers School at Camp Endicott just before the unit was deactivated at Davisville. (Courtesy Betty Summers in memory of Daniel W. Summers.)

Three

CONSTRUCTION BATTALION CENTER DAVISVILLE

This "Seabee," the official symbol of Naval Construction Forces worldwide, was for many years positioned near the main entrance to the Naval Construction Battalion Center, Davisville, just off Route 1 in North Kingstown, Rhode Island. It was fabricated after a design by Frank Iafrate, a civilian worker at Quonset Point Naval Air Station. The "busy bee" carries the tools of the trade and a submachine gun, embracing the true spirit of these Navy men. The name "Seabee"—phonetically identical to CB (short for construction battalion)—was the logical name for this symbol, insignia, and that special group of Navy personnel. Iafrate himself signed on as a Seabee in 1942 and went through boot training at Camp Endicott. (Courtesy Carl A. Passarelli, EAC, retired.)

R.Adm. Alexander "Ace" C. Husband, CEC, USN, (ret.) assumed command of the Naval Construction Battalion Center, Davisville, Rhode Island, in late 1955. He served as Commanding Officer until mid-1958. Admiral Husband was the first and only Commanding Officer of CBC, Davisville to become Chief of Civil Engineers and Chief of the Bureau of Yards and Docks, the duties for which he was sworn in on November 1, 1965. In May 1966, the Bureau was renamed the Naval Facilities Engineering Command. Admiral Husband retired in August 1969 and died in January 1978. (Navy Photograph in Gloria A. Emma Collection.)

This northeasterly aerial view of the Davisville facilities shows Quonset Point in the upper right portion of photograph. The Davisville "triangle" is formed by Newcomb Road (left), Route 1 (front), and the road and rail spur from right center to upper left. The Davisville piers and other significant facilities not in the picture are beyond that margin. (Courtesy Charles E. Carr.)

The Davisville warehouse district can be seen in this aerial view. WWII Camp Endicott was situated just beyond the water tower. The trailer park is also visible in that vicinity. The main rail spur into Davisville is in the foreground. CBC, Davisville had 52 warehouses with a total area of 1,366,000 square feet. Included were five dehumidified buildings. (Navy Seabee Veterans of America, Island X-1, Davisville, Rhode Island.)

This image shows the main gate in September of 1967. That year, the Naval Construction Battalion Center, Davisville, homeported 8,154 personnel and tenants, and employed 1,017 civilians. Capt. John D. Burky, CEC, USN was the Commanding Officer. At the time of the 25th anniversary, the center housed the 21st Naval Construction Regiment, and Mobile Construction Battalions One, Six, Seven, Forty, Fifty-three, Fifty-eight, and Seventy-one. (Collection of Gloria A. Emma.)

The 1960s were filled with challenges and accomplishments at the Seabee Center. Support of the Vietnam effort and Operation Deep Freeze topped the agenda. Morale was high among the Seabees and the civilian employees who supported them; the Chief of the Bureau of Yards and Docks frequently visited Davisville—his former command. In the photograph above, taken on August 21, 1969, R.Adm. Alexander C. Husband reviews his Seabees at an evening parade in his honor. (Navy Photograph by G.W. Gilligan, PH2.)

In 1992, the Davisville Main Gate was at the end of a phase-down cycle begun in 1975 when the center was designated a reserve activity. In the intervening years, the Seabee presence steadily diminished. Gone were the uniformed Gate Guards and the lofty banners of years before. The demise of the Seabee base was to happen 18 short months later. Approaching the main gate for the 50th reunion celebration was a sad moment for all those who had served with and for the Davisville Seabees. (Navy Seabee Veterans of America, Island X-1, Davisville, Rhode Island.)

There was no big marching band at the 50th reunion in 1992, but the Seabees saw to it that their day would be spent with dignity; as a time to remember those who had served before them; and as a symbol of encouragement to those who would follow in their footsteps. Here, members of Reserve Naval Mobile Construction Battalion Twelve hold their flag high as they join with pride in the last and final parade at CBC, Davisville. (Navy Seabee Veterans of America, Island X-1, Davisville, Rhode Island.)

The Administration Building, commonly known as Building 101, dates from WWII. It is configured of five buildings that are connected by a single main corridor on each floor. In the early days of CBC, Davisville, this three-story wood frame structure was the largest office building on the base, claiming 163,857 square feet of floor space. In addition to a complex of administrative offices, Building 101 housed the headquarters offices of the Commander, Naval Construction Battalions, U.S. Atlantic Fleet, a medical dispensary, the dental clinic, the ADP Facility, two auditoriums, and two concrete multi-story vaults. (NORTHDIV, Naval Facilities Engineering Command.)

From left to right, Tony Sarro, Philip Bessette, Doris Lawton, Howard Burdick, and Charles Kelly can be seen busy at work in the Records Management Office of the Administration Department. (Collection of Gloria A. Emma.)

The Administration Department, situated in Building 101, was at the hub of NCBC, Davisville operations. This office was situated in close proximity to the Commanding Officer and was responsible for collecting and analyzing base-wide performance data on behalf of the Commanding Officer. The office was under the supervision of Mr. Edward G. Riley, who was assisted by Mr. Charles J. Kelly, John Kavanagh, Frank Perri, and Gloria A. Emma. (Collection of Gloria A. Emma.)

Louise Allan worked in the Budget Office. Note the typewriters and calculators, as opposed to today's modern office equipment, which includes computers in most offices. (Courtesy Louise Allan.)

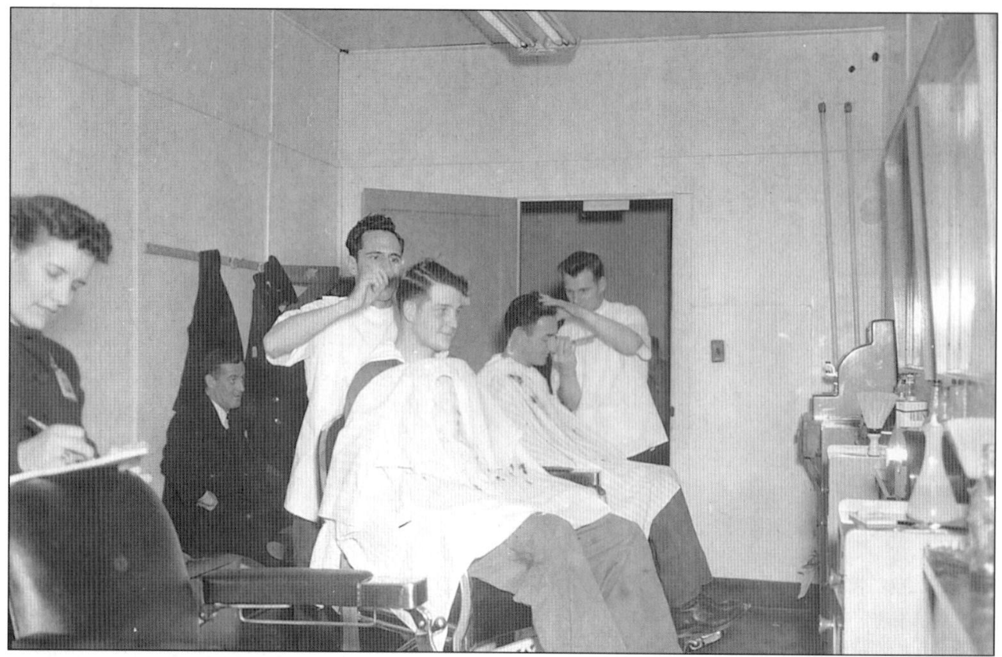

CBC Davisville photographer Virginia Dulleba is shown here, writing a story about the Military Barber Shop, which was located in the far wing of Building 101. (Collection of Virginia Dulleba.)

Here, a Pharmacist Mate is taking a blood sample from a worker. The pharmacy, which was located on the ground floor in Building 101, primarily served the military and their dependents. (Collection of Virginia Dulleba.)

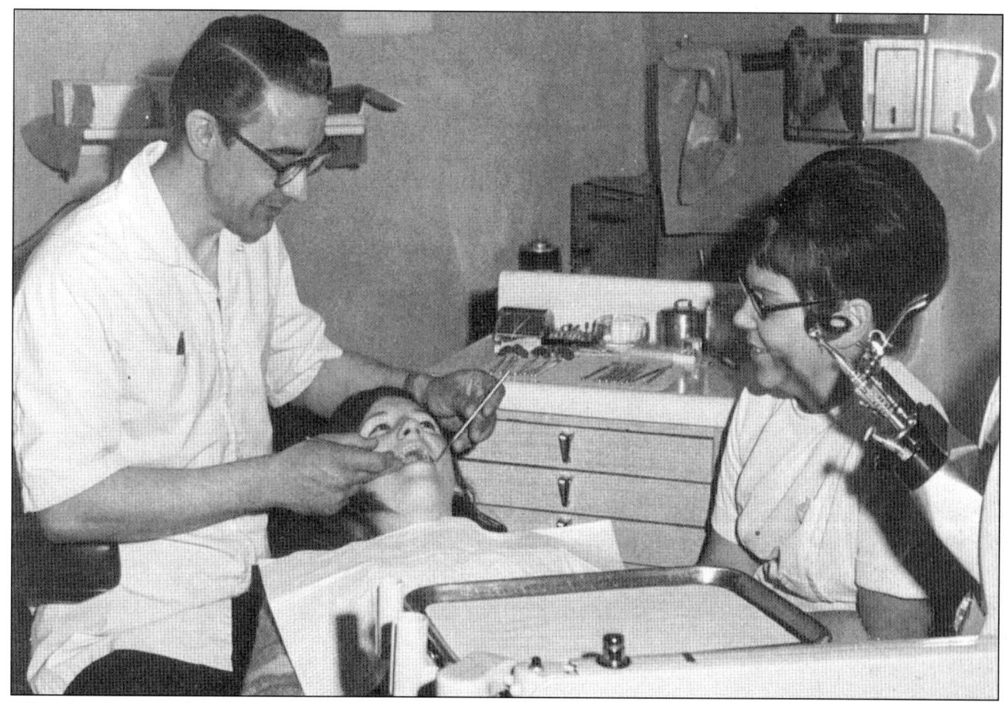

The CBC Dental Laboratory, located on the second floor of Administration Building 101, was headed up by Capt. Louis Pistocco, shown here working on a patient, Kathy Kelly. Looking on is Dental Hygienist Rachel Brouillette. (Welcome Aboard CBC Davisville, 1973.)

The Commander, Naval Construction Battalions, U.S. Atlantic Fleet (COMCBLANT), established his seat in the eastern end of Building 101 on October 1, 1951. CBC, Davisville hosted COMCBLANT and provided logistical support for this tenant. As part of a Shore Establishment Realignment, COMCBLANT was relocated to Naval Amphibious Base, Little Creek, Virginia, in October 1973. (Seabee Memorial Calendar, 1975.)

Building 404 was built to accommodate the 21st Naval Construction Regiment after the wooden structure in which it was originally housed was destroyed by fire in 1968. In the early 1970s, it became the Command Office of Naval Construction Battalion Center, Davisville. Building 101, a temporary WWII–era building, was then vacated and the CBC administrative staff moved to the newer quarters in Building 404. (NORTHDIV, Naval Facilities Engineering Command.)

The Navy's Rapid Deployment Forces, Project Reindeer Station and Operation Deep Freeze, were a few of the projects actively supported by the Naval Facilities Engineering Command Contracts Office in Building 404. Increased workload in the 1980s required expansion of working spaces. In 1986, the unit awarded 215 contracts with a total value of $111.2 million. (Walter K. Schroder photograph.)

A total of nine Enlisted Men's Barracks were built at Davisville after WWII; each had 35 rooms for berthing. These three-story buildings were of concrete construction and had a floor area of 21,702 square feet. (NORTHDIV, Naval Facilities Engineering Command.)

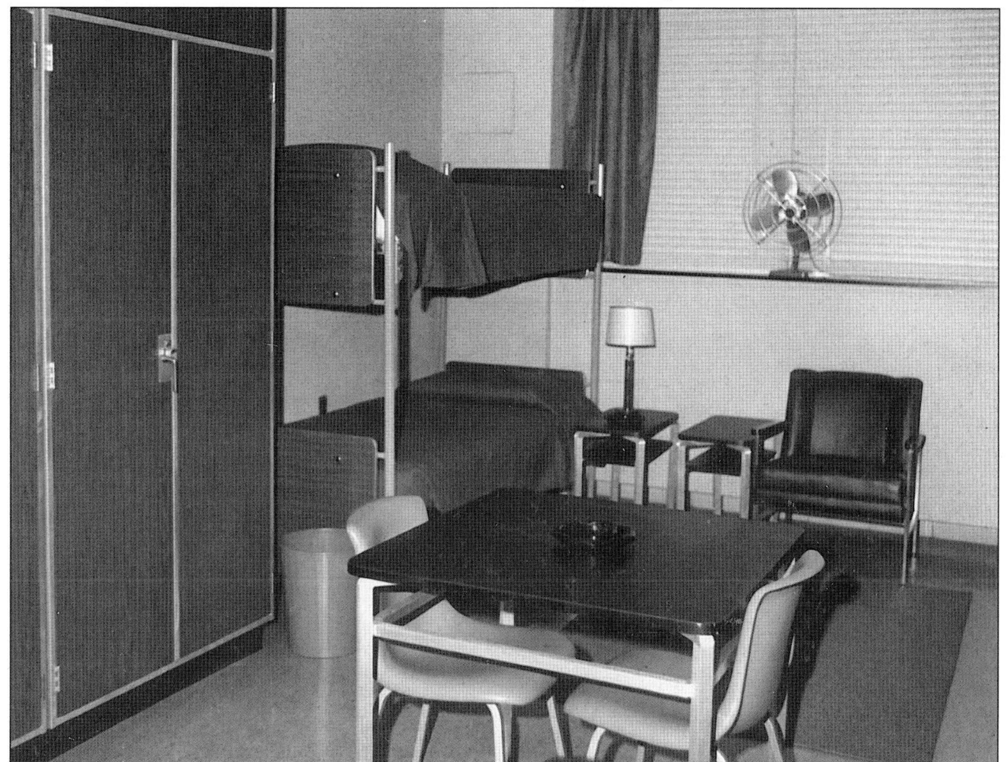

Four-man rooms in the Enlisted Men's Barracks had a floor area of 300 square feet. Barracks and rooms were well furnished and maintained, qualifying them as "Homes away from Home." (Navy Seabee Veterans of America, Island X-1, Davisville, Rhode Island.)

These Officers' Quarters were originally set aside for the Commander Fleet Air Quonset and his family. Eventually, they were assigned to the Commanding Officer, Naval Construction Battalion Center, Davisville. (Rhode Island Economic Development Corporation.)

This is one of several on base Married Officers' Quarters. This building had a floor area of 4,180 square feet, was of wood frame construction, and was heated by an oil-fired boiler. (NORTHDIV, Naval Facilities Engineering Command.)

The first chapel at Davisville, other than an earlier temporary facility at Camp Thomas, was configured from a Quonset Hut, as depicted here. It was used by personnel of all faiths, with services for Catholics and Protestants offered at different times. This chapel was declared surplus in 1947 and moved by the Seabees to the Newport Naval Hospital. (Courtesy Charles J. Kelly.)

The Chapel of the Pines was formally dedicated on September 14, 1963. It was built as a training project for Seabees of five Construction Battalions that were homeported at Davisville at that time. The chapel was the first on-site religious facility at Davisville since an earlier wartime building used for that purpose was declared surplus in 1947. The existing structure is of concrete; when in operation, it seated 132 people. In total, the building has 2,418 square feet of space. (NORDIV, Naval Facilities Engineering Command.)

R.Adm. Alexander "Ace" C. Husband, Chief of Civil Engineers and Commander, Naval Facilities Engineering Command and Capt. John D. Burky, Commanding Officer, CBC, Davisville, participate in the singing of a hymn during a memorial service conducted by Protestant Chaplain Doesman in the Seabee Chapel. (Collection of Gloria A. Emma.)

Mass was celebrated everyday at noontime for the military and civilian personnel during Lenten Season. These daily Masses were very well attended. Clergy from churches in the surrounding area were requested to participate. (Courtesy Louise Allan.)

This marriage of civilian co-workers, which took place on October 10, 1982, was a "first" for the Davisville Chapel in the Pines. Basil DiStefano of the Public Works Department wed Linda Salvagno, Administration Department, in a beautiful ceremony attended by many military and civilian personnel. (Courtesy Mr. and Mrs. Basil DiStefano.)

The happy couple, Mr. and Mrs. Basil DiStefano, emerge from the Chapel in the Pines after giving their vows. They pause momentarily outside the chapel to contemplate the beauty of the moment, their future, and to pose for picture taking. (Courtesy Mr. and Mrs. Basil DiStefano.)

The Enlisted Dining Hall, also known as "The Galley," could seat 428 men and, if needed, 480. It had the capacity of feeding 2,550 men in 2 hours. Total floor area was 23,897 square feet. The dining hall was divided into three dining rooms that were tastefully decorated. (Navy Seabee Veterans of America, Island X-1, Davisville, Rhode Island.)

The Food Service personnel, both military and civilian, were proud of their reputation to be operating one of the best naval messing facilities in New England over a period of years. They enthusiastically pursued a goal they had set for themselves; namely, of pleasing all their customers all of the time. (Welcome Aboard CBC Davisville, 1973.)

The CBC, Davisville Enlisted Dining Hall was rated best in the First Naval District in seven out of nine years. It was rated third best in the U.S. in 1965 and 1970. As is evident in the above photograph, the Food Service Staff was delighted with the results of the competition—they emerged in first place. (Courtesy Nick Cerra.)

Daniel E. Byron served thousands of Davisville Seabees during the 19 years he was Rhode Island Area Director of the United Services Organization (USO). He is shown here with the recipient of the Daniel E. Byron Leadership Trophy. (Courtesy Charles J. Kelly.)

The Seabee Theater, of permanent construction, is capable of seating 500. It was one of a series of on-base facilities frequented by the Seabees that were homeported at Davisville. (Collection of Gloria A. Emma.)

It was not difficult to fill the Seabee Theater to capacity. In addition to offering entertainment for the troops, official ceremonies, concerts, plays, lectures, and other presentations took place at this central facility. In the above photograph, the color guard is about to enter from the rear as one of the first steps in opening an official ceremony. (Collection of Gloria A. Emma.)

The Seabee Concert Band was always on hand to provide the needed rhythm as at this St. Patrick's Day celebration. The music men included all the talent needed to enable the unit to strike up the band at parades, at ceremonies, dances, and concerts. They are known to have performed at the Iwo Jima Memorial in Washington alongside the Seabee Drill Team. (Collection of Virginia Dulleba.)

Professional performers came to entertain at the Seabee Center Theater from New York City. Many officers, Seabees, and civilian personnel enjoyed the performances monthly. (Collection of Gloria A. Emma.)

Kate Smith, a well-liked singer and entertainer of U.S. troops in WWII, sang for the Seabees while visiting Davisville's Camp Endicott in 1943. Here, Frank Iafrate solicits her autograph for a caricature he painted on a wall backstage while she was performing on stage. Kate Smith became popular through her first rendition of "God bless America" in 1938. During the war, she sold over $600 million worth of War Bonds, more than any other person in America. (Courtesy Frank J. Iafrate.)

Frank Iafrate could be found wherever the entertainers were. Here, he is seated while sketching one of his many caricatures. These performers were visiting Camp Endicott during the war as part of an USO entertainment group. They are, from left to right, Catherine Westfield, Akim Tamiroff, Martha Stewart, Leonid Kinskey, and Bill Barlow. (Courtesy Frank J. Iafrate.)

Hollywood actress Celeste Holm sits for one of Frank Iafrate's famous caricatures while on an entertainment tour at Camp Endicott in the 1940s. The actress performed in theater and films. Her greatest fame came from her role in the play *Oklahoma*, performed in 1943. After the war in 1947, Miss Holm garnered an Oscar as a supporting actress in *Gentlemen's Agreement*. In 1956, she starred in *High Society* with Frank Sinatra. (Courtesy Frank J. Iafrate.)

Much of Davisville's blue-collar activity took place in these buildings. The Supply Department and the Construction Equipment Department (CED) shared these spaces, each pursuing its responsibilities in support of assigned Seabee missions. Outside storage of construction equipment can be seen along the upper margin of the photograph. (Navy Seabee Veterans of America, Island X-1, Davisville, Rhode Island.)

A sad day it is looking at these vacant and silent buildings in 1999, where production had been in full swing only a few years ago with workers maintaining and repairing the heavy pieces of construction equipment the Seabees would need in their worldwide deployments. CED had occupied 103,000 square feet of shop and office space and could repair 72 pieces of equipment simultaneously. (Walter K. Schroder photograph.)

CBC, Davisville, and in particular, the Construction Equipment Department, was tasked with storing the so-called Mobilization Reserve Stocks. This entailed the operation of a maintenance and surveillance program for construction equipment that would guarantee its readiness for immediate issue. The Automotive Shop was a part of the overall program charged with maintaining the equipment in "like new" condition. (Welcome Aboard, 1957.)

One of the largest programs carried out by the Construction Equipment Department was for the major repair and overhaul of equipment for the Mobile Construction Battalions in the Atlantic area. Generally, this pertained to equipment returned from overseas for repair. The Heavy Duty Equipment Shop was in the forefront of this program. (Welcome Aboard, 1957.)

The Greasing Shop was a busy place, keeping up with the flow of equipment in the maintenance process. As early as 1957, the Construction Equipment Department performed periodic maintenance on some 800 pieces of equipment permanently assigned to the center. This included automotive, construction, and materials-handling equipment ranging from pick-up trucks to floating cranes. (Welcome Aboard, 1957.)

To maintain and repair all of the heavy equipment shipped to CBC, Davisville for such services, a considerable variety and quantity of parts and tools were required for which the center was accountable. These employees appear to have enjoyed their work at the Tool Bin as can be seen from the expressions on their faces. Thousands of different items had to be stocked and replenished periodically to satisfy the needs of the various shops. (Courtesy Louise Donohue.)

Packing and Preservation was located in Building 319. It was an integral part of the Supply Department operation to fill orders for shipment to Seabee units overseas. Of equal importance was Packing and Crating, a unit responsible for crating the heavy and bulky items destined for Seabee units in Antarctica, at Reindeer Station in the Indian Ocean, and in the hot zones of Vietnam. (Courtesy Nick Cerra.)

The Property Disposal Office and Surplus Sales Office were located in the field. Their responsibilities included the reallocation of items no longer required by the Seabees or the base or sale to the highest bidder. These included, but were not limited to, damaged or usable equipment, vehicles, and office furnishings. Upon the return of the "Deep Freeze" contingent, there were also from time to time live Huskies that had to be dealt with in the normal disposal and bidding process. From left to right, the Property Disposal staff consisted of Mary Imbriglio, supervisor Martie Shinn, Peggy Green, and Claudia Cinquegrana (front). (Courtesy Mary Imbriglio Malcolm.)

The winter of 1978 is no doubt etched in the memories of the center's snow removal crew, which was comprised of employees of the Construction Equipment and Public Works Departments and supported by personnel from the parts supply section of the Supply Department. While others were stranded on the road or at home during the Blizzard of '78, these men made their way to the base to commence clearing operations. Seabees from Gulfport, Mississippi, were flown in to help the city of Providence dig out from the 27-inch snowfall. (Courtesy Luigi Del Ponte.)

The Public Works Department was responsible for keeping the overall plant and the many facilities of the Seabee Center in shipshape. Personnel of this department had steady jobs doing carpentry and plumbing work around the base wherever needed. Whether repairs were required on roads, fences, furnaces, or dripping faucets, the Public Works Department was there to fix it. (Welcome Aboard, 1957.)

The Davisville Piers were especially busy during the Korean and Vietnam Conflicts, when workload and personnel figures reached an all time high. Supplies and equipment measuring 456,025 tons and valued at almost $40 million were transported on 153 ships from these piers to Southeast Asia from April 1965 to May 1971. (NORTHDIV, Naval Facilities Engineering Command.)

The Ammunition Ship USS *Nitro* (AE-23) was berthed at the Davisville Piers when she was not on a run with an underway replenishment group, or off to an advanced base where she served as an afloat ammunition reserve storage depot. The USS *Mazama* (AE-9) also called Davisville her homeport. She was tasked with rearming U.S. aircraft carriers and other ships in the waters off Vietnam. (Welcome Aboard CBC Davisville, 1973.)

The mission of Naval Schools Construction (NAVSCON) assigned to this building was to administer formal courses and special training programs assigned by the Chief of Naval Training. In 1973, NAVSCON offered 15 formal courses plus special training. The school developed and instructed the only dredging course in the Navy. NAVSCON was assigned 31 buildings and had a capacity of 812 students. (Navy Seabee Veterans of America, Island X-1, Davisville, Rhode Island.)

In this 1957 classroom scene, a number of Seabees attend carpentry and construction training at NAVSCON. Basic and apprentice courses were taught, as well as refresher courses, so that Seabee personnel could maintain their technical proficiency while preparing for future deployments. (Welcome Aboard, 1957.)

Seabee Reservist Walter G. Boll, CEP1, can be seen on a lineman training pole in the "First Works Area" in September 1952. This training was conducted just east of the Seabee Theater. (Courtesy Walter G. Boll, CEP1.)

Tool storage facilities were housed in tents within the First Works Area in 1952, until they were replaced by small Butler Buildings that were constructed nearby the following year. (Courtesy Walter G. Boll, CEP1.)

In 1966, both Seabees and civilians received specialist training on multi-fuel engines, which was conducted in the Motor Room of the Construction Equipment Department located in Building 224. (Courtesy Charlie Carr.)

This Seabee work crew was en route to the test area in a line truck in 1954; they needed to set up poles and facilities for lineman testing at that site. (Courtesy Walter G. Boll, CEP1.)

On August 13, 1963, 15 men from Mobile Construction Battalion Eight donned full battle gear and started out on a 50-mile hike from Davisville to Galilee and back. Their packs were heavy, the asphalt hot, and the temperature nothing to rave about. Eight of the Seabees sweated it out all the way; several others were returned by truck. The village of Wickford got a glimpse of a few diehards as they made their way back to the base. (Courtesy Willis Smith.)

Modern-day "Sea Huts" are shown here being erected by members of the 21st Reserve Naval Construction Regiment as a training exercise in the late 1970s. These hands-on projects were conducted in the general vicinity of the former First Works Area. (Courtesy Carl A. Passarelli, EAC, retired.)

Members of the 21st Reserve Naval Construction Regiment were photographed near the "Big Bee" and the main gate to Davisville during their 1991 tour of active duty. The unit was homeported at Davisville from 1978 to 1993, when it was disestablished. (Courtesy Carl A. Passarelli, EAC, retired.)

Seabees ship out by boat, by plane, by bus, or by car; they take whatever means of transportation is appropriate to their assignment and destination. Here, a group of Bees loaded down with their gear board the landing ship *Ashland* at Davisville in June 1966 for a trip to No Man's Land Island for training. (Providence Journal Company photograph.)

Young Sea Cadets were also at home at CBC, Davisville in the 1980s. This crew received training on a U.S. Coast Guard craft in 1985. From time to time, groups would meet at the Seabee Center for boot training. Lcdr. Battle of East Greenwich, along with Comdr. Mario aRussillo from Johnston, was instrumental in establishing the Sea Cadet Training Program at Davisville. (James J. Tringale photograph.)

These Sea Cadets are literally "getting their feet wet" as they go through an early phase of boot training as members of the Davisville Sea Cadet Training unit. They do not appear to have been wearing their Sunday shoes. (Courtesy Walter G. Boll, CEP1.)

Rhode Island citizens joined military and civilian officials to honor the memory of the late Congressman John E. Fogarty on September 23, 1967, when the Sun Valley military combat training area off Route 2 was renamed Camp Fogarty. Sun Valley had been commonly known as a military training area in southern Rhode Island used by Seabees and run by Marines. Mobile Construction Battalion Seventy-one had spent two weeks here a year before preparing for deployment to Vietnam. (Collection of Gloria A. Emma.)

The dedication of Camp Fogarty coincided with the 25th anniversary of CBC, Davisville. Many official dignitaries participated, including Sen. John O. Pastore, Sen. Claiborne Pell, Rep. Fernand J. St. Germain, and Rep. Robert O. Tiernan. Congressman Robert L.F. Sikes from Florida was the principal speaker. Among the ranking military at the dedication were Vice Admiral John T. Hayward, President of the Naval War College, Newport, and R.Adm. Alexander C. Husband, Chief of BuDocks. (Collection of Gloria A. Emma.)

The memorial was designed by Lt. William D. Martin, CEC, USN, Planning and Facilities Engineering Officer for the Seabee Center. A 38-ton boulder, on which the bronze tablet was mounted, had been brought from Arcadia, Rhode Island, by Seabees of the 21st NCR. A concrete foundation and light-colored Beldon paving brick had been placed by MCB Fifty-eight. (Collection of Gloria A. Emma.)

Members of the COMCBLANT Drill Team formed the color guard and presents arms during the Camp Fogarty ceremonies. Camp Fogarty—Sun Valley—was famous for its strict discipline and rigid training schedules. (Courtesy Charles J. Kelly.)

Recommissioned in 1966, MCB Forty was not spared a "grueling" period of training at Camp Fogarty. Here, a detail from Headquarters Company pauses, waiting for what comes next. MCB Forty served in Vietnam, Diego Garcia, Hawaii, the Philippines, Taiwan, and Guam. (Courtesy Steven K. Proctor.)

This Range Observer was perched high above where the training of Seabees at Camp Fogarty was held in order to have an unobstructed view of the firing exercises, enabling him to maintain range safety. A loudspeaker system was at hand for use in giving firing commands and other instructions. (Courtesy Carl A. Passarelli, EAC, retired.)

Rifle practice is the mainstay of any foot soldier. The Seabees were soldiers, in addition to being the Navy's builders. Thus, they had to prepare to trade their shovel or wrench for the small arms at their disposal to defend themselves, if need be. Here, a member of the 21st RNCR gets his chance at Camp Fogarty for some serious target practice. (Courtesy Carl A. Passarelli, EAC, retired.)

Women are now in the Seabee ranks, where they serve as equals, train as equals, and perform as equals. Here, a member of the 21st RNCR takes aim at a target during an active duty training session at Camp Fogarty. (Courtesy Carl A. Passarelli, EAC, retired.)

Off-base military housing was maintained at Navy Drive, off Devil's Foot Road near Route 1. This small enclave consisted of a number of wood frame houses. It included 62 units, 52 comprised of three bedrooms and ten comprised of two bedroom Officers' Housing Units. (NORTHDIV, Naval Facilities Engineering Command.)

Not forgotten was the need for Disaster Recovery Field Training. Davisville maintained such a facility, referred to as "Disaster Village," where Seabees were trained in fire-fighting and rescue operations. The training course area consisted of acreage with buildings and structures. (NORTHDIV, Naval Facilities Engineering Command.)

Four

POST-WAR CHALLENGES

Seabees of MCB Seven board a plane at Quonset Point Naval Air Station in 1964, on their way to Camp LeJeune for training prior to their deployment to Guantanamo Naval Base, Cuba. (Courtesy Capt. A. Robert Socha, CEC, USN, retired.)

In February 1946, just before the 53rd Naval Construction Battalion turned in its equipment in anticipation of being decommissioned, word was received to prepare for shore construction at Bikini Atoll in the South Pacific, the site selected for proposed A-Bomb tests. Bikini Atoll is oval shaped, approximately 20 by 21 miles and consisting of 21 islands. Men of the Fifty-third were assigned to 15 of them. Here, Comdr. John Burky, Commanding Officer, and Comdr. Prothero, Executive Officer, show off the sign to their Pacific paradise. (Courtesy Capt. John D. Burky, CEC, USN, retired.)

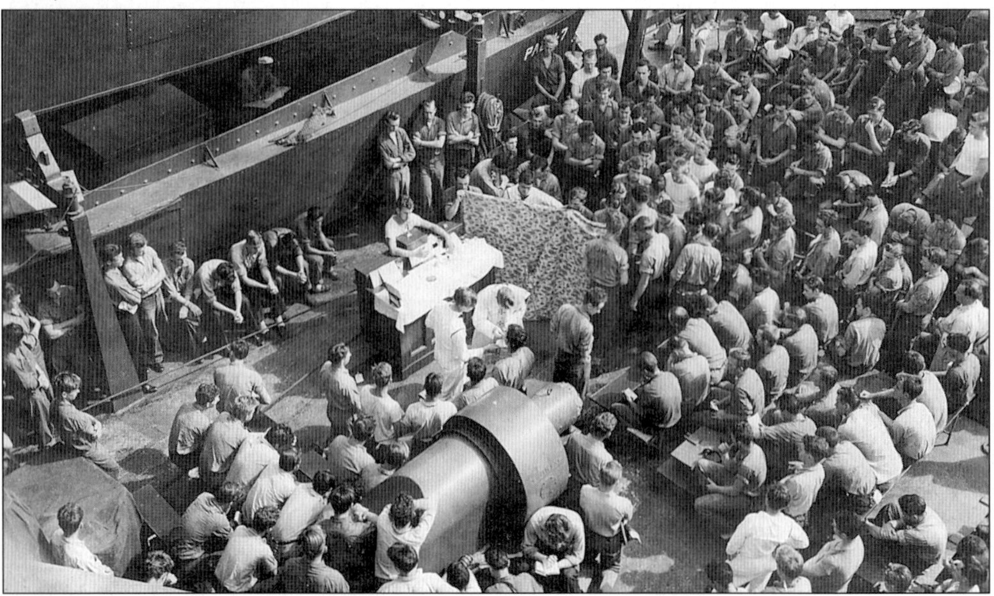

The 53rd NCB can be seen at Sunday religious services en route from Guam to Bikini in March 1946. Most of the old hands had been discharged soon after the war, so the unit was made up of many new Seabees who had little experience. With the guidance of a few good officers, some old-timers, and a real "can do" spirit, the outfit accomplished what it set out to do and more. (Courtesy Capt. John D. Burky, CEC, USN, retired.)

Secretary of the Navy James Forrestall inspected the progress at Bikini Atoll. At the time of the A-Bomb testing, there were approximately 357 ships in the target array, plus 39,000 personnel at Bikini and many neutral observers who lived aboard the ships. The recreational facilities on the island accommodated about 25% of all personnel; as a result, only one out of four was allowed ashore each day. (Courtesy Capt. John D. Burky, CEC, USN, retired.)

A number of camera towers were erected at the test site by the Fifty-third NCB. These were regular Advance Base Control Towers, approximately 100 feet tall; they were beefed up to carry a heavy lead shielded housing for cameras. The towers were guyed to withstand the terrific winds that were expected from the explosion of the A-Bomb. (Courtesy Capt. John D. Burky, CEC, USN, retired.)

Each camera tower was powered by a generator set in an ammunition hut, buried under some 3 feet of earth. Fuel tanks and pumps were also buried to keep the generators running during the tests. (Courtesy Capt. John D. Burky, CEC, USN, retired.)

Although no units were dispatched to Korea from Davisville, the Seabee Center made many shipments of supplies and equipment to that area in support of Seabee forces originally homeported at CBC, Port Hueneme. Individual Davisville Seabees did eventually join the Korean conflict. While in Korea, Robert E. Pruett, SW1, who had served in WWII, shows off the motto he believed in. He later also served in Vietnam. (Courtesy Bobbie J. Todd, daughter.)

These Seabees unload equipment from AKA-55 to a pontoon pier at Wolmi Do, Korea. CBC, Davisville contributed significantly to the flow of supplies to this hostile area. (Courtesy Comdr. John Seites, CEC, USN, retired.)

MCB Four was busy at Guantanamo Bay, Cuba, in 1954, with site preparations for new housing. Gitmo, as it is called, took on a prominent role during the Cuban Missile Crisis. (Courtesy P.R. Murphy.)

By 1955, MCB Four was doing roadwork across the Atlantic at the U.S. Naval Air Station in Kenitra, Morocco. Various Davisville units were assigned to this area over a period of years commencing in 1951. Battalions that built the base were: MCB One, MCB Four, MCB Six, MCB Seven and a detachment from MCB Eight. After 1959, these battalions were sent to Rota, Spain. (Courtesy P.R. Murphy.)

In 1956, MCB Four was assigned to Newfoundland to participate in the construction of a 1,500-man Advance Base Camp at Argentia. Things went just fine until unexpected mishaps slowed the team down. Looks like an unfriendly Bee stung a tire. (Courtesy P.R. Murphy.)

On July 13, 1965, MCB Seven began loading aboard the USNS *Buckner* — destination Rota, Spain. This was the first time since 1960 that MCB Seven had deployed to this great naval base in Southern Spain. The "Fighting Bee" stood proudly at the entrance to the Seabee camp. The letter "E" stands for Excellence, a designation that had been awarded to the unit. (Courtesy Capt. A. Robert Socha, CEC, USN, retired.)

The MCB Seven Drill Team performs at Rota in this photograph. The team was awarded the Navy "E" for excellence, a well-deserved honor. Official recognition of this kind not only made the men feel proud of themselves, their unit, and their performance, but also served to strengthen their morale. (Courtesy Capt. A. Robert Socha, CEC, USN, retired.)

The above collage graphically portrays the varied jobs performed by MCB Seven while deployed to Rota. The thrust of their assignment was the so-called USA Home Project, a task that would produce a group of multi-family duplexes for the dependents of American Service personnel stationed at the U.S. Naval Base in Rota. (Courtesy Capt. A. Robert Socha, CEC, USN, retired.)

In preparation for an administrative inspection, MCB Seven held an exacting greens inspection 60 days after their arrival at Rota. Marine personnel inspected the assembled battalion and quizzed the men on any and all aspects of their weapons. Individual arms as well as assigned recoilless rifles and mortars were given the "once-over." (Courtesy Capt. A.Robert Socha, CEC, USN, retired.)

Guantanamo Naval Base, Cuba, was on the deployment agenda for most Seabees homeported at Davisville. There was a constant rotation of units to this critical station during the Cold War. Here, members of MCB Eight attend an outdoor meeting in the Seabee camp area. (Courtesy Comdr. John Seites, CEC, USN, retired.)

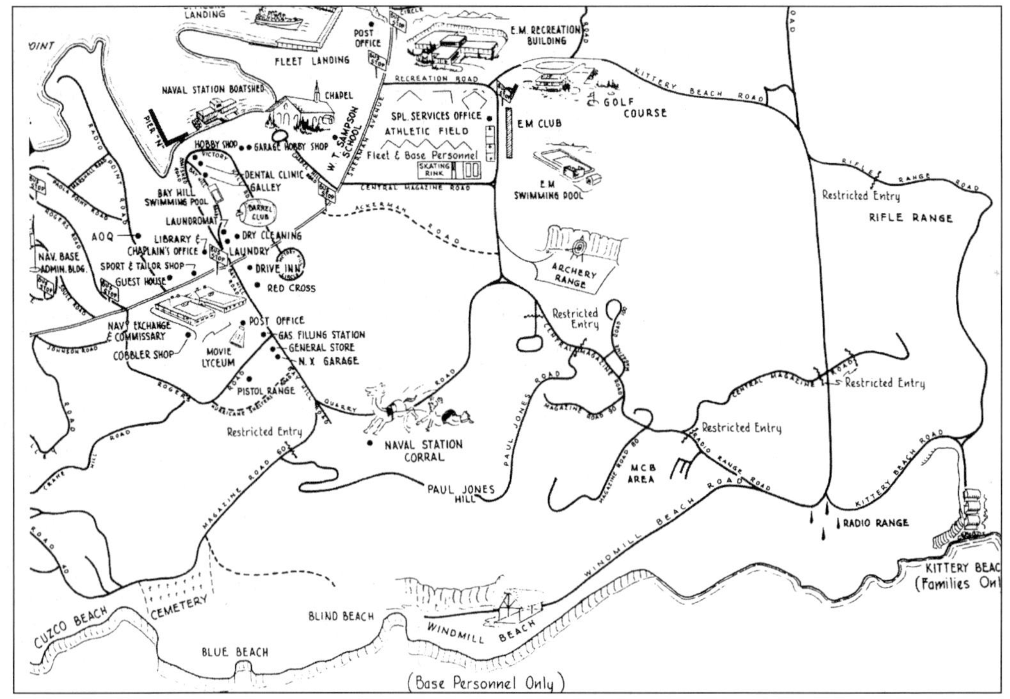

This map of Guantanamo Naval Base, also called GITMO, provides an insight into the size of this naval complex, which was actively supported by Seabees from Davisville, Rhode Island. Most naval facilities were located in the upper section of the map. The Seabees were concentrated in the MCB area at the bottom center of the map. (Courtesy Capt. A.Robert Socha, CEC, USN, retired.)

The Seabee camp at Gitmo provided the necessary housing for the Construction Battalions deployed to this station over the years while they completed specific job assignments in support of the naval station. While there, the Seabees honed their military skills, including compass reading, guerrilla warfare, hand-to-hand combat, and camouflage and personal concealment; these were all necessitated by the prevailing Cold War. (Courtesy Comdr. John Seites, CEC, USN, retired.)

Windmill Beach is a rocky coral-bound stretch of the warm Caribbean, only a few hundred yards from the Seabee Camp at Gitmo. It was perhaps the most popular Seabee weekend recreation spot at this overseas station. Here the "Eightballs," a music group from MCB Eight, entertains their buddies and guests at the beach's big cabana. (Courtesy Comdr. John Seites, CEC, USN, retired.)

The Red Dog Inn was well known to any Seabee who ever served at Guantanamo Bay. It was the favorite watering hole for Seabees after a hard day's work. (Courtesy Lou Demas, EO3.)

Personnel of Operation Deep Freeze '61 sort out baggage upon arrival at McMurdo Sound, Antarctica, during one of the worst spring snowstorms that season. (Navy Photograph by PH3 A.E.Tilley, USN in John Forman Collection.)

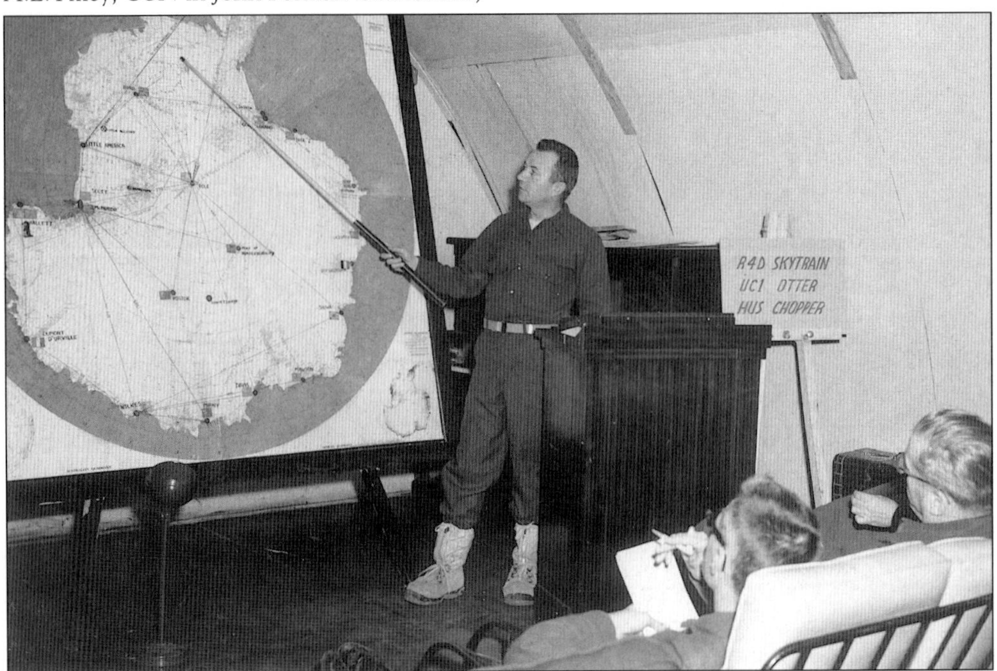

Capt. William H. Munson, Commanding Officer, VX-6 Squadron, briefs the press and guests of R.Adm. David M. Tyree, USN, Deep Freeze Commander, on the numerous missions performed by the squadron in Antarctica. (Navy Photograph by CP Mate F. Kazukaitis, USN in John Forman Collection.)

McMurdo Station, Antarctica, was established on December 20, 1955 by personnel of Naval Mobile Construction Battalion Special during Operation Deep Freeze I. Above is an aerial view of the station in 1960. (Navy Photograph by PHCM C. Hagerty, USN in John Forman Collection.)

Supplies from homeport and support facilities at Davisville, Rhode Island, can be seen after delivery to McMurdo Station, Antarctica, during Deep Freeze '61. Note the stenciling on the front left crate: FROM C.O. NCBC DAV. E. GREENWICH, Rhode Island. Most supplies for this remote station were purchased at NCBC, Davisville, then assembled, packed, and transshipped to Antarctica aboard two annual re-supply ships. (Courtesy Comdr. John Seites, CEC, USN, retired.)

Whenever Seabees arrived in Antarctica, they were greeted by the permanent inhabitants of that cold region. Penguins were always fun to look at and watch; they were friendly too! (Courtesy Willis Smith.)

Seabees and scientists were strangers to Antarctica. They had to dress for survival on this barren continent. Layered clothing, quarters that would protect them from the harsh environment, and an ample cache of food and supplies allowed some of them to "winter-over" at remote stations that were completely cut off from the world until the arrival of another spring. (Courtesy Willis Smith.)

At the bottom of the world, road signs point every which way. Distances in this vast land of permafrost spoil any ideas for weekend travel. There was nothing of any special interest to the Seabees at this lonely place that was less than 8,000 miles away. (Courtesy Willis Smith.)

This chow line is not much different than any other chow line in a Seabee encampment, except these fellows are being served in Antarctica, where they may stay for a prolonged period of time due to the extreme weather conditions. Some of the men will winter over in their quarters under the cover of snow. These Bees are with Task Force Forty-three, participating in Deep Freeze '64. (Courtesy Willis Smith.)

Our Lady of the Snow, a monument erected to the memory of Richard Williams, USN, who died during Operation Deep Freeze I, is examined by members of BVX-6 on Operation Deep Freeze II in 1957. McMurdo Station, Antarctica, is in the background. (Navy Photograph Courtesy John Forman.)

There is no place on earth where the Seabees will not build a chapel, if time permits. This Chapel of the Snows at McMurdo Station, Antarctica, was well frequented by those assigned to this desolate area of snow and ice. (Courtesy Comdr. John Seites, CEC, USN, retired.)

The "Chalet" at McMurdo Station is where the National Science Foundation administers the overall U.S. Antarctic program. The building, procured by the Davisville Contracts Office, was staged and transshipped by the Supply Department on one of two annual re-supply ships to the Antarctic continent. (Courtesy Comdr. John Seites, CEC, USN, retired.)

This sketch of the New South Pole Station shows the dome and the structures within, which were acquired at Davisville under contract and transshipped to Antarctica, where the station was assembled by Seabees homeported in Rhode Island. (Welcome Aboard, 1973.)

This Bachelor Officers Quarters (BOQ) at McMurdo Station, Antarctica, was known as "George's Hotel." The facility, not far from the docking area where all supplies were brought ashore, was used during the Antarctic summer season. Two re-supply ships arrived each year that were unloaded here. Distribution of supplies to more remote stations was made via overland tracked vehicles, or through direct air support. (Courtesy Comdr. John Seites, CEC, USN, retired.)

Seabees of Task Force Forty-three scrape loose snow that has been blown out of the chutes of the Peter Snow Miller and accumulated on the side walls of the main tunnel at Byrd Station, Antarctica, so that the walls will be smoother. (Navy Photograph by PH2 Larry D. Sayan, USN in Collection of John Forman.)

Here, Antarctic Support Activity personnel work from the 40-foot construction bridges to put up 40-foot wonder arches over the trench, which will be the garage tunnel for New Byrd Station, Antarctica. The project was completed in 224 days in 1960. (Navy Photograph by PH2 Larry D. Sayan, USN in Collection of John Forman.)

Construction at Byrd Station, Antarctica, during Deep Freeze '64 included renovations of tunnels and buildings and the erection of an antenna for the upper atmosphere physics program. As can be seen, the Byrd tunnel was used to store supplies that stayed frozen. Note the frost on boxes, pipes, and wires. (Courtesy Willis Smith.)

MCB Four departed Davisville on October 7, 1965, for Vietnam. Here, a young Seabee carrying his rifle and personal gear waves farewell from the front door of the bus, as he begins his trip to the fighting zone in Southeast Asia. These young men left cheerfully and with great confidence, knowing they had been through thorough training and were able to successfully turn their "can do" motto into deeds. (Providence Journal Company photograph.)

MCB One served in Vietnam in 1967 where it upgraded 15 miles of a national highway and constructed over 200 advance base structures near DaNang, while simultaneously repairing and upgrading 30 bridges. Here, members of the Headquarters Company Ready Platoon stand by prepared to perform security duties. (Courtesy R.Adm. Paul R. Gates, CEC, USN, retired.)

MCB Forty had been in Vietnam in 1966. A year later it was back home, and soon after, they went back to Vietnam, where they hauled 75,000 yards of crushed rock needed to produce 60,000 tons of asphalt to pave Route 1, a major national highway. Here, an asphalt paving crew of Alpha Company is busy getting the job done. (Courtesy Steven K. Proctor.)

While some members of MCB Forty were building roads, this crew was fitting a steam pipeline for the camp laundry. (Courtesy Steven K. Proctor.)

Asphalt paving became a specialty of MBC Forty. Here, a bridge over the so-called "Perfume River" gets a special treatment of hot asphalt. (Courtesy Steven K. Proctor.)

During a rocket attack on Camp Campbell, two men from MCB Forty were hospitalized and 14 others treated at a dispensary before being released. Such attacks, in addition to sniper fire, accounted for the award of 17 Purple Hearts. The ceremony in Vietnam is shown here. (Courtesy Steven K. Proctor.)

Awards and ceremonies are a part of military tradition. Here, members of MCB One parade proudly past the reviewing stand during their deployment to Vietnam. The Admiral was on board that day, as his flag tells. (Courtesy R.Adm. Paul R. Gates, CEC, USN, retired.)

This security guard gives a proper hand salute at the entrance to the MCB One camp in the Red Beach, DaNang, area of Vietnam in 1966. He proudly wears the Seabee emblem on his helmet for anyone to identify him as a "can do" Seabee. (Courtesy R.Adm. Paul R. Gates, CEC, USN, retired.)

Commander R.K. White, CEC, USN, who assumed command of MCB Forty in Vietnam during the closing weeks of their second deployment, returned to Davisville with his men on completion of that assignment. Welcoming them home at Quonset Naval Air Station is Capt. C.C. Heid, CEC, USN, Commanding Officer of the Naval Construction Battalion Center, Davisville, along with other dignitaries and friends. (Courtesy Steven K. Proctor.)

The arrival of MCB Forty at Quonset Point called for the presence of an Honor Guard and the presentation of arms to each and every man, as a sign of respect and gratitude for their service in Vietnam. Capt. D.G.Iselin, CEC, USN, Commander, Naval Construction Battalions, U.S. Atlantic Fleet is the first in line to greet the homecomers. (Courtesy Steven K. Proctor.)

For a period of four years, commencing in March 1970, CBC, Davisville was involved in the planning, support, and construction of a Naval Communications Station on the Island of Diego Garcia in the Indian Ocean. Assembled at the airstrip on July 19, 1971, from left to right, are Lcdr. Rao, SC; Marcel Mouline, Plantations Manager; Lt. Col. McNamara, the first pilot to land at Diego Garcia; and Comdr. Dan Urish, C.O. of MCB Forty. (Courtesy Comdr. Daniel W. Urish, CEC, USN, retired.)

Diego Garcia became the most unusual and largest peacetime effort for the Naval Construction Forces. Davisville personnel, both civilian and military, had a hand in every aspect of what was known as Project Reindeer Station. Here Comdr. Phil Oliver, C.O. of MCB One, inspects the Seabee encampment with Comdr. Dan Urish, C.O. of MCB Forty, before assuming command in November 1971. (Courtesy Comdr. Daniel W. Urish, CEC, USN, retired.)

Flags fly over Reindeer Station on the Island of Diego Garcia in August 1971. The new Naval Communications Station, completed in 1974, was situated in the Indian Ocean, some 11,000 miles from Davisville. (Courtesy Comdr. Daniel W. Urish, CEC, USN, retired.)

Stored at Davisville for a number of years were rusty components of a prototype AMMI dry-dock, which utilized pontoons as pier and lift sections that were flooded with water to submerge. The dock could accommodate a 2,500-ton destroyer. Detail Zulu from MCB Seventy-one had assembled the dry-dock near the Davisville piers in 1969. (Courtesy Luigi Del Ponte.)

Needs for lighterage to accompany the Navy's Rapid Deployment Forces during the height of the Cold War resulted in refurbishing the AMMI-dock components stored at CBC, Davisville. (Courtesy Luigi Del Ponte.)

The size of the various AMMI-dock sections becomes clear when viewing the movement of the completed components to the docking facilities via the runway of Quonset Point Naval Air Station. (Courtesy Luigi Del Ponte.)

Seabees displayed with pride their CAN DO motto wherever they went. Car tags as this insured their story would reach into all corners of America. (Collection of Gloria A. Emma.)

The Big Bee was at the main gate, the Little Bee close to the entrance to Building #404. (Walter K. Schroder photo.)

Five
Home Port Moments

Just as Davisville Seabees participated in civic projects overseas to the benefit of the civilian population, so they raised funds for worthy causes at home in Rhode Island. Here Frank Iafrate, designer of the famous Seabee emblem, demonstrates his talents during WWII by entertaining an audience during a fund-raiser at Roger Williams Park, Providence, Rhode Island. (Courtesy Frank J. Iafrate.)

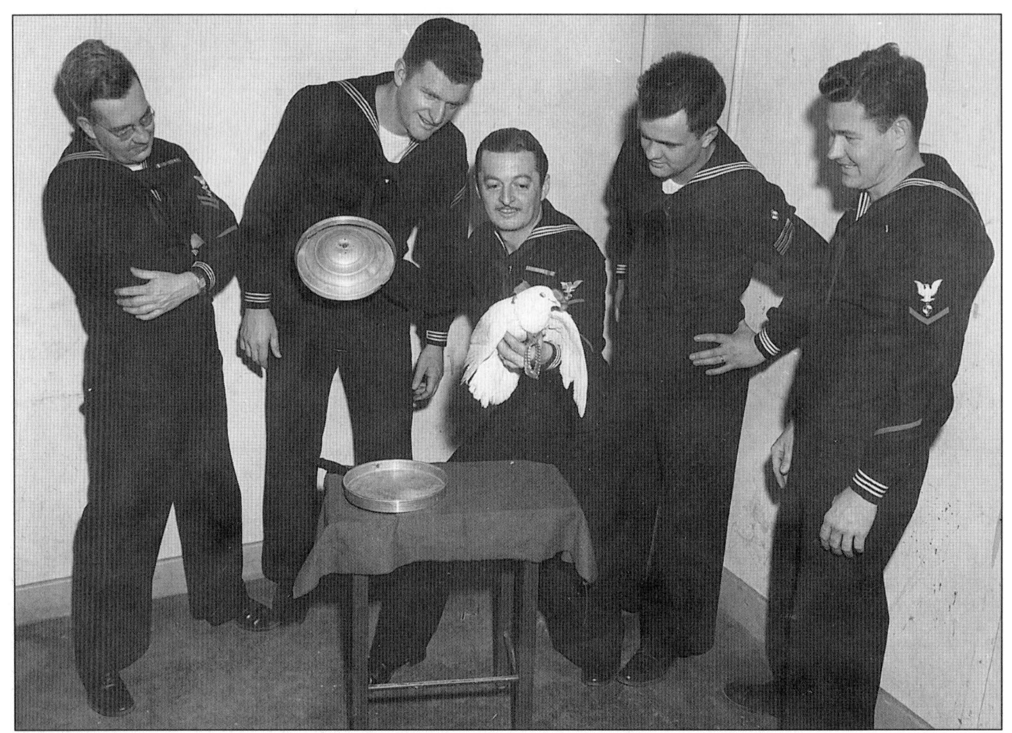

Never wavering, the Davisville Seabees insisted they "can do" any job. If the normal routine of hard work did not produce the desired results quickly, they would simply do magic. They always succeeded. This group of Seabees are being entertained by one of their own while homeported at Davisville during WWII.(Courtesy Frank J. Iafrate.)

From time to time and when available, the Ceremonial Band from MCB Seventy-one entertained Davisville civilian workers at brown bag lunch concerts. The civilians found the stirring music inspiring and the Seabees got in some extra training while enhancing their overall public image. (Courtesy Dennis Jansson.)

Civilian workers Terry Murphy and Ruth Frank enjoy the company of their favorite Seabee, commonly known as "Ski," who was assigned to Davisville as a Military Personnel Technician. They are seen here in front of the huge ship's bell that stood near the entrance to the Administration Building for many years. (Collection of Gloria A. Emma.)

Mr. Edward G. Riley, Administrative Officer in the '50s, came up with the idea of a baseball game to be played by female employees of the base. One team was called the "Bloomer Gals" and the other was named the "Dogpatch Dolls." Of course, the Bloomer Gals won. The game was held at lunchtime and all military and civilian personnel turned out for it. The "hat" was passed around for donations and all the money collected was routed to the United Fund. This was only one of the many events at Davisville that benefited the United Fund. (Collection of Gloria A. Emma.)

To everyone's surprise, a mystery guest showed up among the Bloomer Gals. It was none other than their Seabee friend Ski, who worked in the Military Personnel Office at Davisville at that time. (Collection of Gloria A. Emma.)

These were the Dogpatch Dolls, recruited from among the female employees of Davisville. They put on a good show, as well as a good fight against the costumed Bloomer Gals, and were cheered for their part in the fund-raising for the United Fund. (Collection of Gloria A. Emma.)

R.Adm. Lewis B. Combs, CEC, USN, and guest of honor at the 1978 Seabee Ball, is shown here addressing over 200 guests. Seabee Balls were held every year since the inception of the base on March 5 in celebration of the birth of the U.S. Navy Seabees. These events were attended by active, retired, and reserve military, CBC employees, and friends of the center. The last Davisville Seabee Ball was held in 1984. (Navy photograph by PH2 Ernest A. Myette, USNR, in Gloria A. Emma Collection.)

Davisville-Quonset Point Navy Cycle Club members and senior officers in the area pose with 24 new mini-bikes that were made available to Navy dependents under a safety-oriented youth motorcycling program during the latter part of 1972. Mounting bikes in the front row, from left to right, are Capt. Warren G. Stevens, Commanding Officer, Naval Construction Battalion Center, Davisville; R.Adm. Joseph B. Tibbets, Commander, Fleet Air Quonset; and Capt. Donald K. Forbes, USN, Commanding Officer, NAS, Quonset Point, Rhode Island. Equipment Operator Second John Morgan, vice-president of the Davisville-Quonset Point Navy Cycle Club and organizer of the mini-bike program, kneels at center. (Courtesy Capt. Warren G. Stevens, CEC, USN, retired.)

On July 23, 1974, the Huddersfield Brass Band from England was aboard CBC, Davisville to entertain military personnel and civilian workers. Here, members of the band perform solo parts at an outside concert. Pat Iaciofano of the Public Works Department coordinated the band's activities during their visit. (Courtesy Pasco Iaciofano.)

The Seabee Marching Band, ready to "sound off," assembled just outside the main gate in front of the center's "Seabee," which is perched on one of the Davisville pontoons of WWII fame. (Courtesy Dennis Jansson.)

Kenneth Lyons, Commander of the Federal Employees Veterans Association, and Capt. William Wesanen distribute monetary awards to civilian employees for beneficial suggestions. (Courtesy Carmine "Gabby" Rivera.)

Women co-workers from within the center attend a wedding shower for Frances Waters during their lunch break. Fran is the girl wearing glasses to the left in the photograph, admiring some of the shower gifts. (Collection of Gloria A. Emma.)

A handshake seals the agreement between heads of the National Association of Government Employees, previously known as the Federal Employees Veterans Association, and the Commanding Officer. Those at the signing, from left to right, were Joe Booth, Civilian Personnel Officer; Walter Pankiewicz, Steward; Capt. John Burky, Commanding Officer; Enrico Chini, Steward; Thomas Hartshorn, President, NAGE; and Jack Martone. (Courtesy Thomas Hartshorn.)

From time to time, NAVFAC sponsored Industrial Relations Field Institutes at selected sites. In this instance, Davisville was chosen to host the conference. Participants from all over the U.S. are shown here enjoying lunch in the CBC, Davisville Galley. (Courtesy Charles J. Kelly.)

Camp Endicott had Frank Iafrate as its favorite illustrator; the NCBC, Davisville looked to Nick Cerra to design posters for any special occasion, and a whole array of different announcements in between. (Courtesy Nick Cerra.)

Members of the Officers' Wives Club display king-sized Christmas cards at the Seabee Center as part of a fund-raising event to support the Wickford Kindergarten. The event netted $150. Here, Chairperson Harriet Jackson (left) presents the cards to teacher Jean Hanna. The cards were made by Nick Cerra, the CBC illustrator. (Courtesy Nick Cerra.)

The Seabee Boosters Committee was active from 1956 until December 1974. The volunteer committee assisted incoming military personnel in finding suitable living accommodations, helping in many ways to integrate them into the community. The Boosters were well appreciated for their dockside services, where they served free coffee and doughnuts to incoming or departing Seabees. (Courtesy Alan French.)

The NCBC, Davisville put considerable effort into supporting the United Fund Campaign. Barbecues and similar events were held annually to support this worthy cause. Here, civilian organizers show off their posters at the Builders Club in 1969. (Courtesy Nick Cerra.)

Decorating the Christmas tree in the Command Office was a yearly ritual. Showing off their decorating skills are, from left to right, Comdr. Bill Wynne, Executive Officer; Charles Kelly, Administration Officer; Capt. Charles Heid, Commanding Officer; Gloria Emma, CO's Secretary; and Carole Adams, XO's Secretary. On a later assignment, Captain Heid was promoted to Rear Admiral. (Collection of Gloria A. Emma.)

The Commanding Officer, Capt. Paul R. Gates, receives his Christmas gift from Santa Claus (Sam Forleo, a civilian employee) at one of the annual Christmas dinners held in the Max Kiel Memorial Gymnasium. Captain Gates was promoted to the rank of Rear Admiral in 1978. (Courtesy Ralph Catallozzi.)

The Duke and Duchess of Windsor visited Davisville during the war years. They are shown here with their host and hostess, Capt. Fred F. Rogers, CEC, USN, Commanding Officer at Camp Endicott, and Mrs. Rogers. Also shown is R.Adm. Gaylord Church, CEC, USN. (Collection of Virginia Dulleba.)

Capt. Bob Wooding, Commander Naval Construction Battalions, U.S. Atlantic Fleet; John H. Chafee, Governor of Rhode Island; and Capt. Joe Barker, Commanding Officer, CBC, Davisville were photographed at a meeting in the COMCBLANT Headquarters Office, located at the far end of the Administration Building #101. (Collection of Gloria A. Emma.)

During one of his many visits to the Construction Battalion Center, Sen. John O. Pastore is shown here viewing a recently acquired machine gun. From left to right are Capt. Robert Schepers, CO CBC; Comdr. Arthur Gardiner, XO CBC; Charles Kelly, Administration Officer, CBC; and an unidentified Seabee. This visit occurred in 1961. (Courtesy Charles J. Kelly.)

At this evening parade in 1969, there was little worry that the base might be phased down in the foreseeable future. All seemed well, but the budgeteers thought otherwise. The Davisville Seabee Center was entering a period of serious realignments over the next few years that would cut into its capabilities to continue its accustomed level of service. (Navy photograph by R.A.Schweitzer in Gloria A. Emma Collection.)

At this conference chaired by Captain Husband, Commanding Officer, CBC, Davisville in the late 1950s, the emphasis was on management responsibilities and the challenges of "Yesterday and Today." The "Tomorrow" was omitted. It could not be readily predicted as it depended on many factors beyond local control. (Collection of Gloria A. Emma.)

Monuments memorializing the past were displayed at the entrance to Building 404 until the deactivation of the center in 1994. These and several other monuments to the memory of WWII Seabee units will be kept in safekeeping until an appropriate site is found to display them with deserved dignity. (Walter K. Schroder photograph.)